101 Outer Space Projects for the Evil Genius

Evil Genius Series

101 Outer Space Projects for the Evil Genius

DAVE PROCHNOW

McGraw-Hill
New York   Chicago   San Francisco   Lisbon
London   Madrid   Mexico City   Milan   New Delhi
San Juan   Seoul   Singapore   Sydney   Toronto

Copyright © 2007 by The McGraw-Hill Companies, Inc. All rights reserved. Printed in the United States of America. Except as permitted under the United States Copyright Act of 1976, no part of this publication may be reproduced or distributed in any form or by any means, or stored in a database or retrieval system, without the prior written permission of the publisher.

1 2 3 4 5 6 7 8 9 0   QPD/QPD   0 1 3 2 1 0 9 8 7

ISBN-13: 978-0-07-148548-7
ISBN-10: 0-07-148548-1

Sponsoring Editor: Judy Bass
Production Supervisor: Pamela A. Pelton
Editing Supervisor: Stephen M. Smith
Project Manager: Patricia Wallenburg
Copy Editor: Marcia Baker
Proofreader: Paul Tyler
Indexer: Karin Arrigoni
Art Director, Cover: Anthony Landi
Composition: TypeWriting

Printed and bound by Quebecor/Dubuque.

This book is printed on acid-free paper.

McGraw-Hill books are available at special quantity discounts to use as premiums and sales promotions, or for use in corporate training programs. For more information, please write to the Director of Special Sales, McGraw-Hill Professional, Two Penn Plaza, New York, NY 10121-2298. Or contact your local bookstore.

Before you attempt to do any of the Evil Genius projects described in this book, please read, understand, and accept the following warnings, precautions, and disclaimers regarding the disassembly of electronics. Thank you.

Precautions

Disassembling electronic devices will void your warranty. There is no authorization for the disassembly or modification of any equipment. There could be a risk of electrical shock or fire by disassembling electronics equipment.

Warnings

Monitors and LCD screens contain dangerous, high-voltage parts. Always remove the battery and disconnect any power cord(s) prior to disassembling *any* electronics equipment.

Disclaimers

The McGraw-Hill Companies and Dave Prochnow will neither assume nor be held liable for any damage caused to anyone or anything associated with the disassembly, modification, and hacking of any gadget, gizmo, or electronics equipment. The warranty for this equipment will be considered null and void if any associated warranty seal has been altered, defaced, or removed.

About the Author

Dave Prochnow is an award-winning professional writer, editor, and contributor to numerous technical publications including *MAKE*, *Nuts and Volts*, and *SERVO Magazine*. He is the author of 27 nonfiction book titles for Addison-Wesley, F&W Publications, McGraw-Hill, and TAB Books, including the bestselling *PSP Hacks, Mods, and Expansions* and *The Official Robosapien Hacker's Guide* (both published by McGraw-Hill). Currently, Dave serves as the Contributing Editor for *Nuts and Volts* and *SERVO Magazine*. In 2001, he won the Maggie Award for writing the best how-to article in a consumer magazine. To learn more about Dave's books and other projects, visit his Web site: www.pco2go.com.

Contents

Contents

Acknowledgments

The amount of support and contributions provided by corporate sponsors during the development of this book was as large as outer space itself. I would like to take this opportunity to recognize and acknowledge the tremendous, unselfish participation by the following people who helped to shape this book into such a wonderful experience for you, the reader.

Jason Baumgarth, Brand Development Director at Zhumell Sport Optics, who kept my eye on the lens; Melly Bonita, Public Relations Director with Konus USA; Nicola Chan at JoinMax Digital Tech. Ltd.; Daniel Da Pont, Sales & Marketing Director for William Optics Co., Ltd.; Staci Dolgin-Rubinstein of the Salmon Borre Group, who fell for my "Giant Ant Farm" joke; Michael Goodman from Imaginova-Orion-StarryNight, who convinced me that "newer" is "better" (Doh!); Jeff Hineline of Great Red Spot Astronomy Products, who is also an instructor at the University of Phoenix and Wayne Country Community College, both in Detroit, Michigan; Angela Linsey-Jackson at Magellan Navigation, Inc.; and Dennis Phillips from The Walker Agency and Barbara Mellman Skinner, both of Bushnell Outdoor Products, who literally helped "pry" a product out of a customer's hands; and, finally, Tricia Richardson, the Marketing Manager with First Texas Products, who was refreshingly patient while "shedding some light" on night vision goggles. Thank you all for your kind and helpful support.

Acknowledgments

Introduction

Space Is the Place

Before I became a world-famous nonfiction author (no, really, this guy in Hong Kong knows me and he thinks I'm pretty good at what I do), I tried my typewriter at spinning fictional yarns. Lucky for me (and you) this creative writing venture didn't pan out.

During this short "fictional" period of my life, I wrote two screenplays and one serialized teleplay. For those of you who really care to learn more about this financially unrewarding point in my career, let me give you a brief overview of each of these dismal failures.

First, there was the serialized teleplay. It was supposed to be a specific genre of TV comedy—situation comedy, both figuratively and literally, and spiced with witty wordplay.

The teleplay was pitched to the one television network that I felt would most benefit from my wit. So much for wit and comedy. It was met with complete disdain from the assembled network program planners.

Well, judge for yourself. I titled the teleplay, *Meal Ticket*. Here, excerpted for your pain or pleasure, is the proposed summary for *Meal Ticket*, as well as a witty scene sample:

Meal Ticket Summary:

One of the three mandatory daily activities that every human being must partake in is eating. In fact, most of us indulge (or overindulge) ourselves with three helpings of this daily regimen. Interestingly enough, there is a curious offshoot phenomenon that occurs during eating—conversation. Even when it is devoid of the customary trappings of lavish locations, stunning sets, and compelling costumes, conversation can be vivid, engaging, and even entertaining. *Meal Ticket* explores the dining-as-entertainment concept through the mealtime conversations of four "regular" people. In each episode, the cast partakes in a meal at various restaurant venues in modern New York City and, we, the audience, get to tag along and eavesdrop on their conversation.

MEAL TICKET: The Pilot

1. ACT ONE

   SCENE A

   INT. RESTAURANT—EVENING

   (LIONEL, REG, PEN, KURT, WAITER)

   LIONEL, REG, PEN, AND KURT HAVE JUST BEEN SEATED AT THEIR TABLE AND LIONEL, REG, AND KURT HAVE BEGUN LOOKING AT THE MENU. THE RESTAURANT IS BUSY AND PEN IS DISTRACTED BY THE ACTIVITY.

REG

   What happens if you type "google" into Google?

PEN

   (Breath) Oh my gosh, is that ...?

KURT

World chaos.

LIONEL

(Looking) Who, where?

PEN

That looks just like. . . .

REG

Does Homeland Security know about that?

(CONTINUED)

WOW, that rocks, doesn't it? No? Well, don't worry, you're not alone in your dislike. The network programmers agreed with your viewpoint and I saw myself to the door. Thank you.

Not dismayed by this rejection (OK, so I was a little dejected), I sat down and knocked out an epic screenplay. Who needs television, anyway?

And *KUNK* was born.

KUNK was a sci-fi/horror genre thriller with a strong undercurrent of campy humor. Derived from "Bigfoot" and "Abominable Snowman" sightings, *KUNK* was a child of atomic bomb testing gone wrong … horribly, but humorously wrong.

The script was tight, but somewhat lengthy due to way too much wordplay between an enormous and disjointed cast. No matter; without fanfare, I shipped copies of the *KUNK* script to every major and minor film studio. And what I received back was a big, fat zero response rate. Not even a "Thank you for your submission, but unfortunately your script doesn't fit in with any of our past, current, or future production plans" kiss-off rejection letter. Nothing. Nada. Null.

OK, I can see you scratching your head now. What do all of these detours down memory lane have to do with outer space and being an evil genius? Well, so far, nothing. Please indulge my

ramblings just a little longer and we will journey out into space. I promise.

So back to the typewriter, for one more script. This one's got to be it. You know, the third one's the charm. The result of this last-ditch final effort was *Seeds*.

This screenplay was pure science fiction. Gone were the failed attempts at humor; instead, only hardcore barebones sci-fi scenes and dialog filled the 100 or so pages of *Seeds*.

Set in a present-day Earth (circa 1979), *Seeds* begins with a government that is attempting to rekindle the golden age of exploration, where average men and women struggled to carve out a niche in the unexplored portions of our planet.

Rather than seeking to chart the mountains, plains, and oceans of Earth, however, the explorers in *Seeds* are teamed up, loaded into space capsules, and blasted into outer space.

Gleaned from millions of wannabe adventurer applicants, the government draws a short list of 100 "couples" who will be admitted into the Space Exploration and Emigration Dispersal Service or SEEDS. Under the pretense of "merging adventure with science," these SEEDS cadets are trained as astronauts who will pilot an Earth spaceship to a distant star and look for inhabitable planets.

What is unknown to everyone in *Seeds* is that the government's reason for forming SEEDS is not solely limited to "adventure and science." In a backchannel research project, the government has learned that an enormous rogue asteroid is less than one year away from smacking into Earth. Likewise, the theoretical calculations drawn from this collision indicate that Earth will be pulverized into several tons' worth of dust from this impact. Yeah, talk about a deep impact.

So, SEEDS cadets are paired as male and female astronaut crewmembers, sealed inside space capsules, planted atop huge rockets, and launched from various missile bases throughout the world—each crew destined for a different distant speck of light in the night's sky—like seeds sown on a prevailing wind of hope.

The remainder of the script examines the fates of several SEEDS crews.

When *Seeds* was completed, I pitched the script to several different film studios, but the results were identical to my previous two fictional writing attempts—rejected.

Rejected, but not downtrodden.

So here I sit, a nonfiction technical "how-to" book author with a whole boatload of great outer space projects for you, the evil genius. Let's get to it.

Space, the Final ... Answer

At first blush, you might wonder who cares about space? Well, look around yourself, because almost everyone has a special interest in the celestial comings and goings that orbit around our little blue marble.

It's in every newspaper you pick up, every news program you watch, and every RSS news feed you read—space, astronomy, planets, NASA, and even science fiction dominate our daily news.

Do you need some proof? Just look at these headlines from a couple of months' worth of news:

- "Virgin Galactic to Build Spaceport in New Mexico," AP, London, UK; December 13, 2005.
- "NASA Capsule Returns First Comet Samples to Earth," AP, Dugway Proving Ground, UT; January 15, 2006.

- "Geologist Walter Alvarez Honored for Dinosaur-Extinction Impact Theory," AP, Reno, NV; March 7, 2006.
- "Scientists Find 'Super-Earth' 9,000 Light-Years Away," by Lewis Smith, The Times; March 14, 2006.
- "European Space Probe Goes Into Orbit Around Venus," AP, Darmstadt, Germany; April 12, 2006.
- "Russians Plan to Put Man on Moon," AP, Moscow, Russia; April 12, 2006.
- "UK: No Visits From Little Green Men," Reuters, London, England; May 8, 2006.
- "States Starry-Eyed Over Spaceports," AP, Los Angeles, CA; May 16, 2006.
- "Japan Offers Free Rocket Shots," AP, Tokyo, Japan; May 18, 2006.
- "Three New Planets Found Around Sun-Like Star," by Ker Than, SPACE.com; May 18, 2006.
- "Former Nazi Removed From Space Hall of Fame," AP, Alamogordo, NM; May 19, 2006.
- "Voyager II Detects Solar System's Edge," by Ker Than, SPACE.com; May 23, 2006.
- "Ancient Rock Art May Depict Exploding Star," by Ker Than, SPACE.com; June 5, 2006.
- "Ancient Rock Carving May Have Recorded Supernova," AP, Phoenix, AZ; June 7, 2006.
- "Hawking: Space Key to Human Survival," AP, Hong Kong, China; June 14, 2006.
- "Hubble Camera Comes Back to Life," AP, Baltimore, MD; June 30, 2006.
- "Shuttle Team Shoos Away Vultures," by Marsha Walton, CNN; June 30, 2006.
- "Large Asteroid Zips Past Earth," AP, Los Angeles, CA; July 3, 2006.
- "'Starshade' May Aid Space Exploration," Reuters, London, England; July 5, 2006.
- "Meteorite Mystery Vexes NASA Scientists," AP; August 7, 2006.

- "Physicist James Van Allen, Discoverer of Radiation Belts, Dies at 91," AP, Iowa City, Iowa; August 9, 2006.

- "NASA Loses Track of Original Moon Landing Tapes," AP, Washington, DC; August 16, 2006.

- "Proposal Would Increase Planets From 9 to 12," AP, Prague, Czech Republic; August 16, 2006.

- "Teachers, Toymakers, Planetariums Await Solar System Overhaul," AP, Washington, DC; August 17, 2006

- "Pluto Faces Possible Planetary Pink Slip," AP, Prague, Czech Republic; August 24, 2006.

- "Giant Puffy Planet Discovered 450 Light-Years from Earth," AP, Washington, DC; September 15, 2006.

- "New Ring Spotted Around Saturn," Reuters, Washington DC; September 20, 2006.

- "Experts Rethinking How Stars Explode," Reuters, Washington DC; September 21, 2006.

- "Astronaut Collapses During Ceremony," AP, Houston, TX; September 22, 2006.

- "Face on Mars Gets Makeover," by Robert Roy Britt, SPACE.com; September 22, 2006.

- "China Cracks Open Door on Secretive Space Program," AP, Cape Canaveral, FL; September 25, 2006.

- "Iranian Women Applaud Space Tourist," AP, Tehran, Iran; September 25, 2006.

- "Rocket Crashes After Launch from New Mexico Spaceport," AP, Upham, NM; September 25, 2006.

- "Japan Launches Sun Satellite," Reuters, Tokyo, Japan; September 25, 2006.

- "Jupiter's Smaller Spot Getting Redder," AP, Washington DC; October 11, 2006.

Now, how about making some headlines of our own? Evil genius style.

To Infinity and Beyond

Divided into 11 chapters, this book covers all the major topics that pertain to astronomy and space exploration. Navigation, rocketry, telescopes, astrophotography, star gazing, space exploration, and, even, alien life forms are each examined inside these 11 chapters.

To better illustrate these astronomy and space exploration topics, a select collection of projects is assembled in each chapter. Want to build your own rocket, telescope, space station, or planetarium? Or, would you like to learn how to determine the altitude of a rocket's flight, create your own star catalog, or take great photographs of the moon? Each of these topics is covered in the 101 projects collected in this book.

Remarkably, all the pieces and parts needed for catapulting your evil genius mind into outer space can be found at your local hobby, craft, hardware, and toy stores. You won't need to make any mad long-distance shopping sprees to ACME Incorporated headquarters.

In fact, each project begins with a complete list of the actual products you will need for achieving success in your evil ways. Unlike the books from those "other publishers," the projects in this book have been painstakingly researched for the best product or products that can meet or exceed the goals and expectations defined for each project.

Therefore, whether you're building a rocket fighter (Project 11) or programming a Mars rover robot's brain (Project 77), you will find a list of all the "preferred" and "approved" materials needed for successfully building the maniacal device described in that project's descriptive title. This list is called "What You Need."

Supporting each list of materials is a helpful compilation of links and references that are relevant to each project. This compilation is called "Resources" and one is at the end of each project.

There's only so much you can do with a bucket of parts, however. You've got to know how to put them together, too.

And what better way to guide you along each project than to actually *show* you how to build it. In a unique presentation technique, enormous photograph "spreads" are used to illustrate each step needed for getting from concept to completed project.

You won't be guessing what comes next or saying, "What did he mean by that?" Just sit back, follow along with each extensive photographic step-by-step monologue, and, before you know it, you'll be exclaiming, "EGADS!"

More specifically, you will be an elite member of the Evil Genius Aeronautics Directorate for Space or EGADS (see Figure I-1).

Of course, no evil project would be complete without an evil genius assistant. No, you won't find Igor in these projects, but you will find the unsavory Penelope (see Figure I-2).

Don't let her looks fool you. Packed into this pint-sized parcel is a diabolical mastermind. Luckily, she tempers her fanatical quest for galactic supremacy with a twist of humor and a

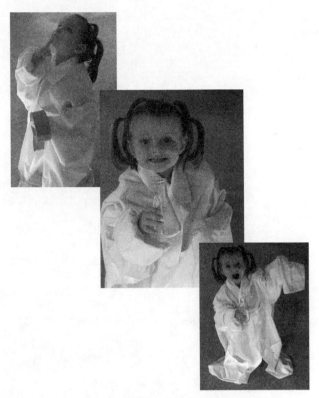

Figure I–2 *Every evil genius needs an assistant. Lacking an Igor, I used Penelope. And she is mad, I tell you, mad. Moo-ha-ha!*

pinch of wit. So, when you need a break from some of the science stuff, you can count on Penelope to dish up some lighter fare.

And she promises not to make fun of Uranus— too much.

So cue the lightning bolt, sound the thunder clap, light your fuse, and let's blast off. Worlds await us, the universe is our oyster; moo-ha-ha.

Resources:

- American Astronomical Society—www.aas.org/
- The Astronomical Journal— www.journals.uchicago.edu/AJ/
- Astronomy & Astrophysics—www.aanda.org/
- European Southern Observatory— www.eso.org/
- ICARUS, International Journal of Solar System Studies—icarus.cornell.edu/

Figure I–1 *The Evil Genius Aeronautics Directorate for Space.*

Chapter One

Let's Go Where No Evil Genius Has Gone Before

Where are you? Right now, at this very moment, do you know your exact position? In today's pushbutton world of digital electronics, global positioning system (GPS) manufacturers like Magellan Navigation® have built a business based on answering these types of questions.

You hear a lot about GPS, but what makes this modern mapping system tick?

What began as a satellite-based U.S. Department of Defense navigation network called NAVSTAR is today known as GPS. Initially launched in 1978, the entire GPS network was completed with its current population of 24 satellites in 1994. Unlike your cell-phone satellite network, however, GPS operates as a free system that can be accessed by anyone anytime anywhere.

Each of these GPS satellites orbits the Earth at an altitude of 12,000 miles, twice a day. While the satellite transmitter is actually powered by a photovoltaic array, tweaks in orbit are accomplished with a set of rocket engines. These adjustments in orbit prevent the satellites from crashing into each other, ensuring that each transmitter operates properly throughout its anticipated ten-year lifespan.

Emitting two signals, known as L1 and L2, that have a transmission strength of about 50 watts, each satellite broadcasts enough information to help you find yourself—literally. Although unable to accurately transmit into buildings, GPS receivers like those from Magellan can help you track yourself down to within 15 meters anywhere in the world.

Although each GPS satellite transmits two signals, known as L1 and L2, only the L1 signal is used by commercial GPS receivers. Operating at a UHF frequency of 1575.42 MHz, the L1 signal consists of three pieces of data:

- **Satellite ID Data**—identifies which satellite your GPS receiver is using for positional information.
- **Ephemeris Data**—orbital information for every GPS satellite.
- **Almanac Data**—satellite status, current date, and time.

If you're looking for greater precision than 15-meter positional accuracy, a Magellan GPS receiver equipped with the Wide Area Augmentation System (WAAS) will supply much better resolution. By using WAAS, you can find yourself within 3 meters of your real position.

To plot your current latitude and longitude position (that is, 2-D position), a GPS receiver needs the signals from three different satellites. More advanced 3-D positional measurements (for example, latitude, longitude, and altitude) require the reception of signals from at least four satellites. Once this data is available, advanced GPS receivers can manage the data into calculations for your speed, heading, and cumulative distance traveled.

These 2-D and 3-D positional placements are determined by triangulation of a signal's transmission and reception times as calculated from the required three or four different satellites. The difference in these two times provides a distance measurement for the movement of each respective satellite. Armed with the distances for at least three different satellites, the GPS receiver triangulates the distances and plots your calculated position on a map.

Project 1. How to Take a Heading

What You Need:

- Silva compass (for example, Base Plate models: Starter™ 1-2-3, Polaris® 177, Explorer™ 203)

Resources:

- How to Use a Compass—www.learn-orienteering.org
- Silva USA—www.silvausa.com
- Suunto Compasses—www.suunto.com

Figure 1–1 *In any mapmaking exercise, there's no substitute for a good compass.*

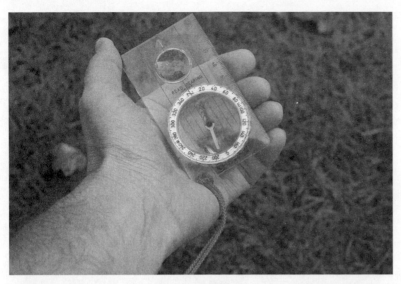

Figure 1-2 *Adjust your compass to find north.*

Figure 1-3 *Accurate headings can be determined by turning the wheel of the compass while retaining the previously determined north setting.*

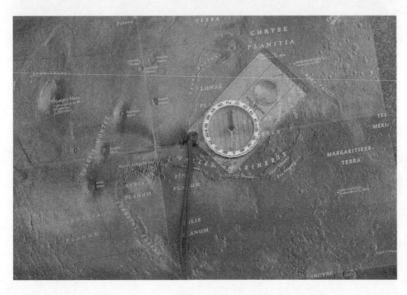

Figure 1-4 *Lay your compass on a map to gain a proper orientation and heading.*

Figure 1–5 *Locate a significant point of interest (POI), and then determine your heading for reaching that POI.*

Figure 1–6 *Some geologic maps are marked with lines that indicate true and magnetic north. Use these lines to make corrections in your heading.*

Figure 1–7 *Use your compass's scale for measuring the distance for your plotted course.*

Project 2. Make Your Own Maps

What You Need:

- Paper (yeah, if you can find any of this antique recording medium)
- Pencil
- Tape measure
- Compass

Resources:

- International Orienteering Federation— www.orienteering.org

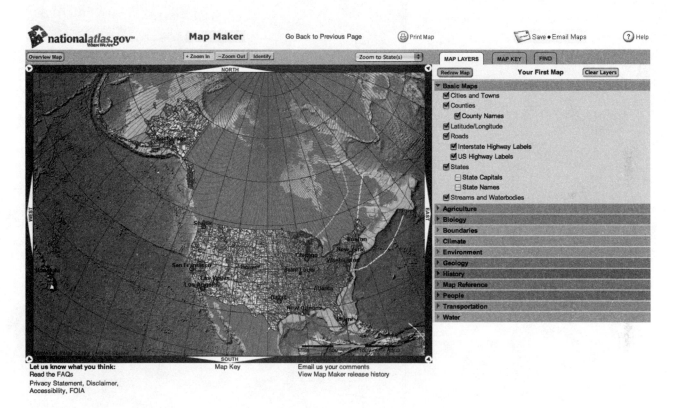

Figure 2–1 *Map Maker from the National Atlas website (nationalatlas.gov) displays lines of magnetic variation.*

What You Need:

- Protractor
- Thread
- Large metal nut or washer
- Tape

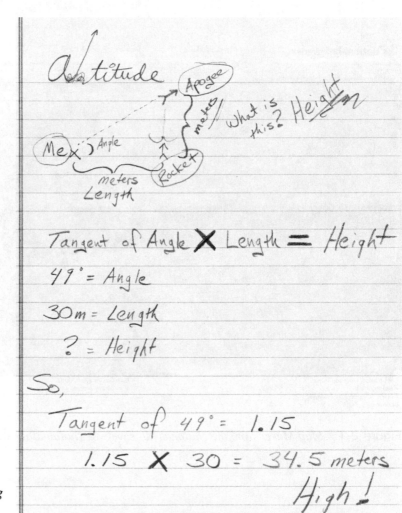

Figure 3-1 *The formula for calculating altitude.*

Figure 3–2 *A simple protractor, a weight, some thread, and a little tape can build a device for measuring the angle used in calculating altitude.*

Figure 3–3 *Attach the weight to the thread.*

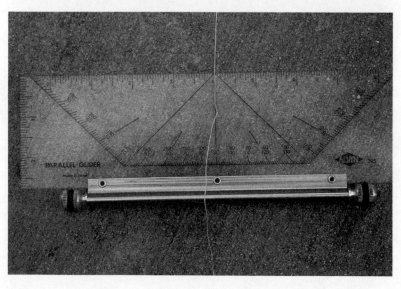

Figure 3–4 *Align the threaded weight with the center of the protractor.*

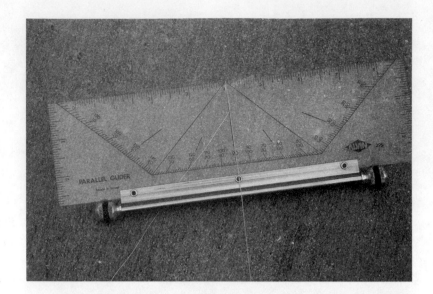

Figure 3–5 *Tape the thread in place.*

Figure 3–6 *As you tilt the protractor, the thread indicates your viewing angle.*

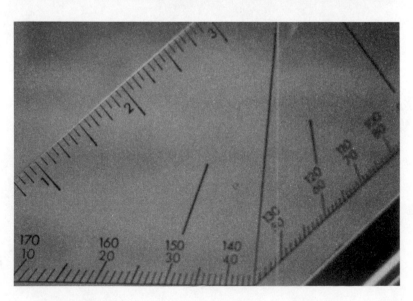

Figure 3–7 *This result indicates a 49° viewing angle.*

Project 4. How to Determine Speed

What You Need:

- Stopwatch
- An accurate altitude reading (see Project 3)

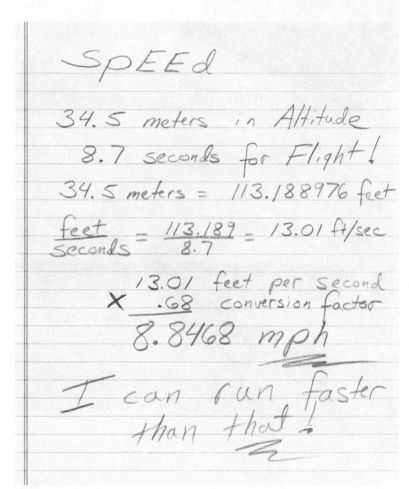

SpEEd

34.5 meters in Altitude

8.7 seconds for Flight!

34.5 meters = 113.188976 feet

$\frac{feet}{seconds} = \frac{113.189}{8.7} = 13.01$ ft/sec

   13.01 feet per second
× .68 conversion factor
8.8468 mph

I can run faster than that!

Figure 4-1   *The formula for calculating speed.*

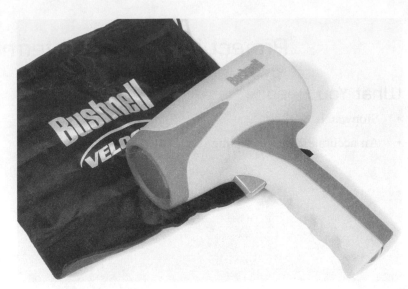

Figure 4-2 *Calculating speed is old school. Speed guns, such as the model by Bushnell, are new school.*

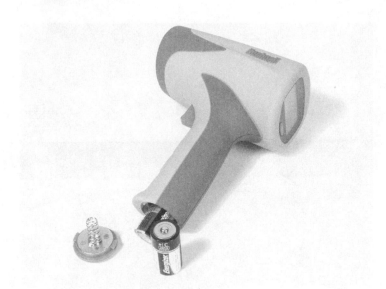

Figure 4-3 *Just load 'er up with a couple of batteries.*

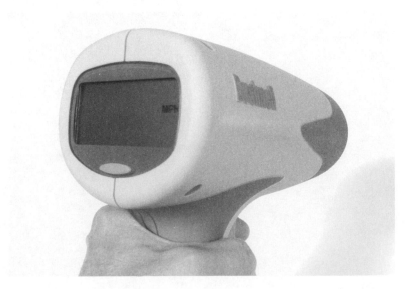

Figure 4-4 *Squeezing the trigger gives you a velocity reading in miles per hour.*

Figure 4–5 *Tiny, fast-moving targets like model rockets can be difficult to measure.*

Figure 4–6 *Here's a great setup for measuring speed and altitude.*

Project 5. How to Determine Acceleration

What You Need:

- Calculator
- An accurate altitude reading (see Project 3)
- An accurate velocity reading (see Project 4)

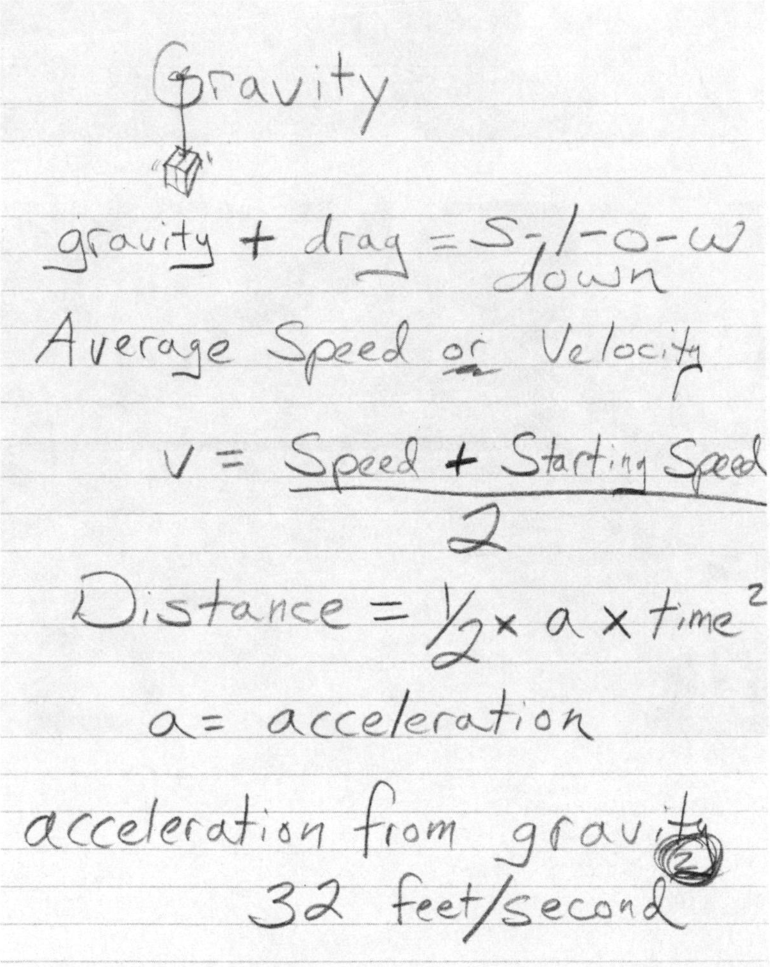
Gravity
gravity + drag = S-l-o-w down
Average Speed or Velocity
v = Speed + Starting Speed / 2
Distance = 1/2 × a × time 2
a = acceleration
acceleration from gravity
32 feet/second 2

Figure 5–1

A couple of ways to calculate acceleration and the effect of gravity on a moving object.

Project 6. How to Use Google™ Earth

What You Need:

- Google Earth

Resources:

- Google Earth—earth.google.com

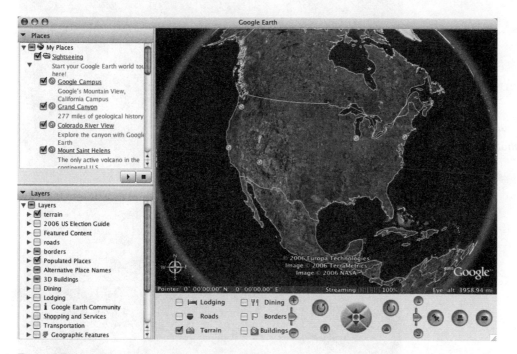

Figure 6–1 *Google™ Earth is a great tool for quickly surveying a vast database of satellite imagery.*

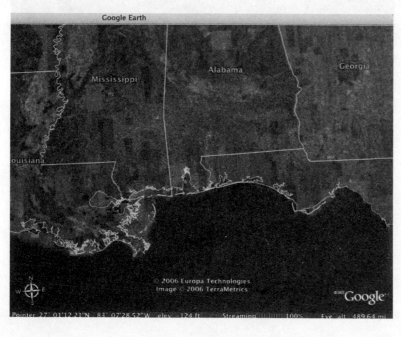

Figure 6–2 *As you maneuver the pointer, Google Earth generates detailed compass heading, latitude, longitude, elevation, and satellite altitude information along the bottom of the browser window.*

Figure 6–3 *You can quickly obtain detailed satellite maps that can be printed for field use.*

Figure 6–4 *You can add points of interest and even spy on your neighbors.*

Project 7. How to Get Found When You're Lost

What You Need:

- Compass
- Map

Resources:

- Earth Science Information Centers—geography.usgs.gov/esic/esic_index.html
- National Atlas of the United States, March 5, 2003—nationalatlas.gov
- U.S. Geological Survey—www.usgs.gov
- USGS Digital Raster Graphics—topomaps.usgs.gov/drg/

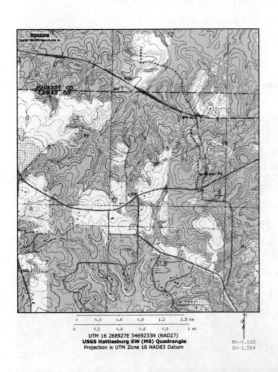

UTM 16 268927E 3469233N (NAD27)
USGS Hattiesburg SW (MS) Quadrangle
Projection is UTM Zone 16 NAD83 Datum

Figure 7–1 *A topographic map from TopoZone shows the same view from Google Earth in Figure 6-4 in Project 6.*

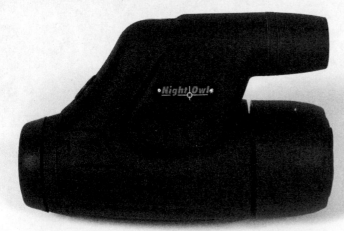

Figure 7–2 *Using a Night Owl Optics® Night Vision scope is essential for finding your way around in the dark.*

Project 8. How to Know the Lat/Long of It

What You Need:

- Topographic map
- Satellite imagery

Resources:

- MacDEM—www.treeswallow.com/macdem
- MapMart—www.mapmart.com
- Maps a la carte, Inc.—www.topozone.com
- Maptech—mapserver.maptech.com
- MICRODEM—www.usna.edu/Users/oceano/pguth/website/microdem.htm
- Microsoft TerraServer Imagery—terraserver-usa .com

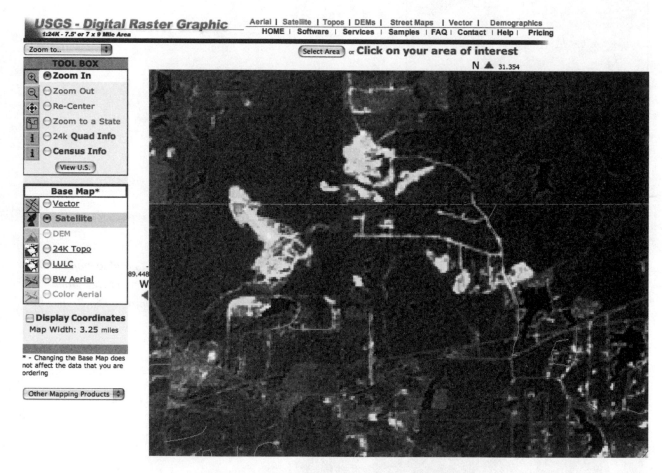

USGS - Digital Raster Graphic
1:24K - 7.5' or 7 x 9 Mile Area

Aerial | Satellite | Topos | DEMs | Street Maps | Vector | Demographics
HOME | Software | Services | Samples | FAQ | Contact | Help | Pricing

Zoom to..

Select Area or **Click on your area of interest**

N ▲ 31.354

TOOL BOX
- ⊙ Zoom In
- ○ Zoom Out
- ○ Re-Center
- ○ Zoom to a State
- ○ 24k Quad Info
- ○ Census Info

View U.S.

Base Map*
- ○ Vector
- ⊙ Satellite
- ○ DEM
- ○ 24K Topo
- ○ LULC
- ○ BW Aerial
- ○ Color Aerial

☐ Display Coordinates
Map Width: 3.25 miles

* - Changing the Base Map does not affect the data that you are ordering

Other Mapping Products

89.448
W

Figure 8–1 *A United States Geological Survey (USGS) Digital Raster Graphic map of the same area in Figure 6-4 in Project 6.*

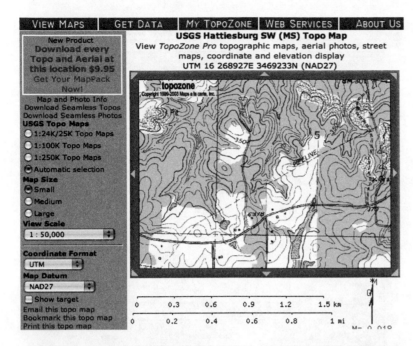

Figure 8–2   *A TopoZone USGS topographic map of the same area in Figure 6-4 in Project 6.*

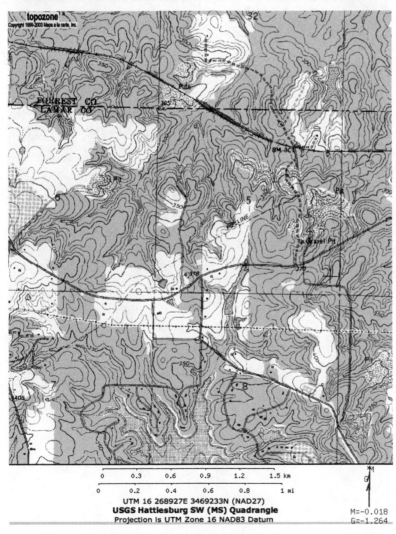

Figure 8–3   *A printed USGS topographic map from TopoZone of this same area near Hattiesburg, Mississippi.*

Figure 8–4 *Not all maps are created equal. This is a Maptech® product for the same area shown in Figure 8-3.*

Figure 8–5 *Maptech maps can also be generated as aeronautical charts.*

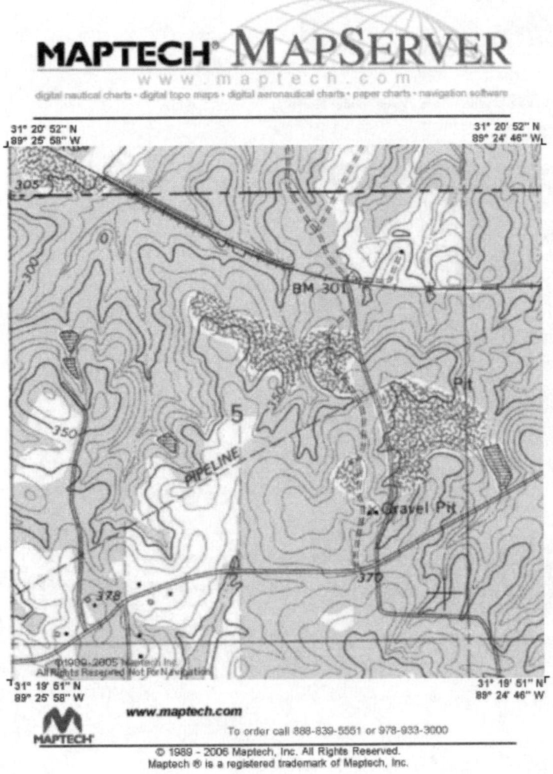

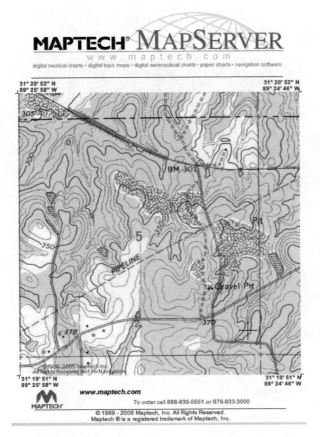

Figure 8–6   *The online Maptech map for our same Hattiesburg sample area.*

Figure 8–7   *A printed sample of a Maptech map.*

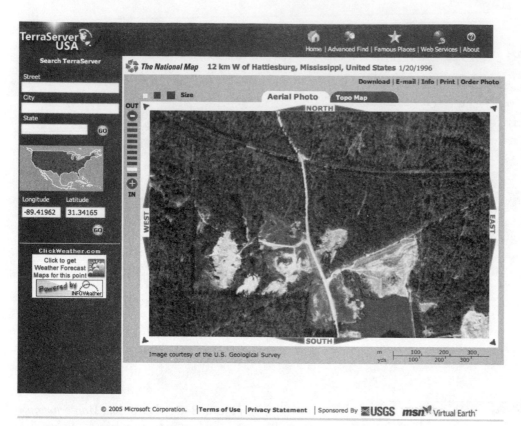

Figure 8–8   *Microsoft Corporation TerraServer.*

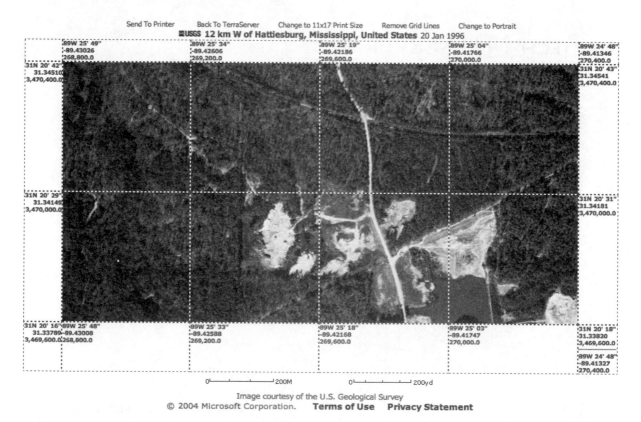

Figure 8–9   *TerraServer generates its maps from the USGS database.*

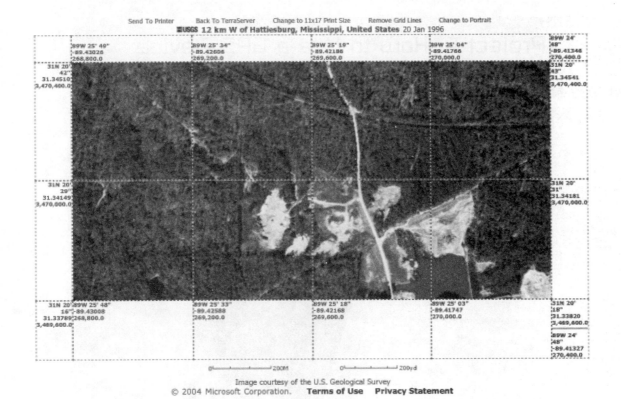

Figure 8–10   *A sample print from TerraServer of our sample area near Hattiesburg.*

What You Need:

- Magellan eXplorist 210

Resources:

- Magellan Navigation, Inc.—
www.magellangps.com

Figure 9–1 *The Magellan eXplorist 210 is ideal for evil geniuses.*

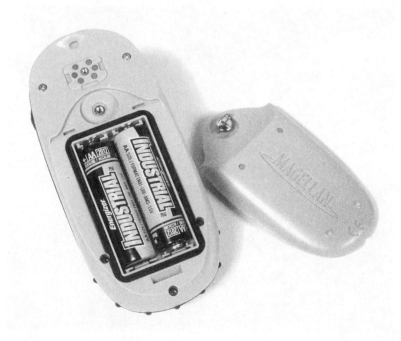

Figure 9–2 *Just plop a couple of AA batteries into the eXplorist 210 and you can have the world in the palm of your hand. Well, at least you'll know where you are in the world.*

Project 9. How to Use a GPS Device

Figure 9–3   *Push the ON/OFF button to start the eXplorist 210.*

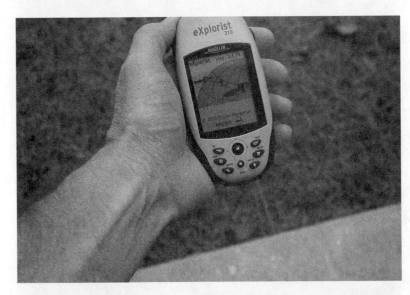

Figure 9–4   *The first step is to help the GPS device find out where you are currently standing.*

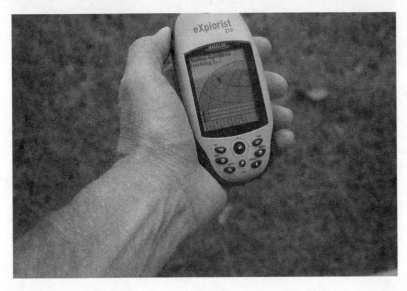

Figure 9–5   *The eXplorist 210 begins to search for satellites.*

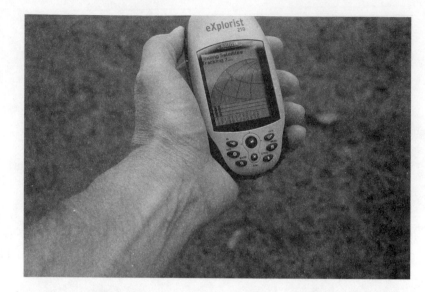

Figure 9–6 *Two satellites have been located. Yes, you could now start using the GPS device, but for best results, you should wait until the eXplorist 210 indicates it has made a "3-D Fix" for your current position.*

Figure 9–7 *You are here. Well, at least I am here.*

Figure 9–8 *You can zoom in on your present location with the Magnify button.*

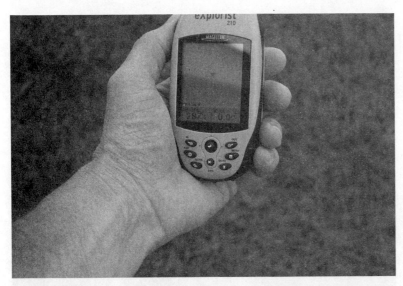

Figure 9-9 *In some parts of the world, it's possible for you to zoom in too close. In this case, you won't have any map reference points to help you gather your bearings.*

Figure 9-10 *A digital compass is built into the eXplorist 210.*

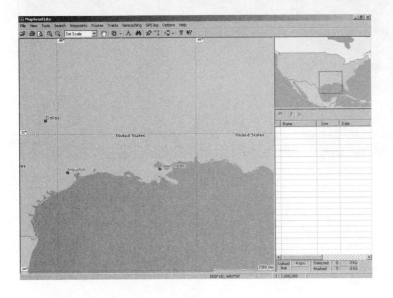

Figure 9-11 *MapSend Lite is an included PC-only program that can help manage your eXplorist 210 maps, routes, and points of interest.*

Project 10. Waypoints, Nav Points, and End Points

What You Need:

- Magellan eXplorist 210

Resources:

- Magellan Navigation, Inc.— www.magellangps.com

Figure 10–1 *A GPS device can be useful for helping you set up your telescope.*

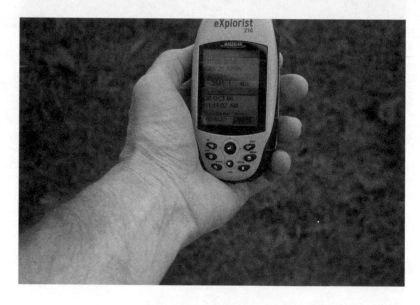

Figure 10–2 *Just about everything you need to know for gaining a full understanding of where you fit into the universe can be determined with a professional GPS device, such as the eXplorist 210. OK, well, almost everything.*

Chapter Two

3–2–1, Blast Off

Learning how to build your own model rocket in the early 1960s would invariably lead you to a quirky book published by Ballantine Books. *Rocket Manual for Amateurs* by (Capt.) Bertrand R. Brinley (1960) was an amazing, albeit ridiculously dangerous, book for budding rocket builders.

Liquid-fueled motors, reinforced concrete test bunkers, and advanced metal nosecones, bodies, and fins characterized the bulk of this handbook's "how to" guidance. Written under the auspices of the U.S. Army (hoping to inspire future generations of Robert Goddard physicists), Brinley's book was packed with dream-inspiring illustrations penned by Barbara Remington. Likewise, an inspirational Foreword by Willy Ley almost compelled you to run into the garage and blow a couple of fingers off your hand.

Model rocketry has come a long way in 40-some odd years. Today, you can purchase safe commercial rocket motors in just about any hobby or craft store throughout the U.S. Pioneering companies, such as Estes and Centuri, revolutionized model rocketry with the production of disposable solid-propellant rocket motors.

Not all model rocketry is based on these solid-propellant rocket motors, however. Alternatives, such as the cosmetically similar Rapier motors and the venerable metal-tubed Jetex (or Jet-X) motors, can both be used for powering your evil attempts to explore the stratosphere.

A small Czechoslovakian disposable rocket engine marketed as Rapier is the ideal power plant for making scale-model rocket gliders. Manufactured in four levels of thrust (L-1 through L-4), the Rapier is a solid-fuel rocket motor that is started with a fuse. Once lit, the Rapier quickly builds up thrust for a burn duration time of 20–25 seconds.

The conventional design scheme for Rapier-powered scale models is to equip your craft with a motor trough. In other words, the rocket motor hangs from the bottom of the aircraft model inside a hollowed-out cavity, which is obscured by the fuselage's side profile. This design element facilitates the rapid insertion and subsequent exchange of Rapier motors before and after each flight.

Although somewhat of a departure from true scale, the trough is an elegant option to the alternative of using intake/thrust augmentation tubing and mounting the rocket motor inside the fuselage. In this case, access to the motor would necessitate the building of a removable hatch. Likewise, adequate cooling for the motor during flight would be paramount for preventing damage to the aircraft.

In most rocket-powered aircraft designs, reusable Jetex motors can be substituted for the Czech-made Rapier motor. One benefit from electing to use Jetex motors is that you won't need to construct a motor mount—a paper and wire container that is essential for holding Rapier motors during flight. The Jetex motor is equipped with a metal clip, which is capable of holding the motor in flight.

Regardless, of whether you use Rapier or Jetex power, you still need to incorporate the motor trough system into your design. Likewise, make sure that you strive to balance your Jetex- and Rapier-powered designs along a center of gravity point that travels roughly through the center of the wing chord.

If the fuse and burn motors don't satisfy your insane plans for world domination, there are other "kinder and gentler" propulsion methods for model rockets. Water, vinegar, and antacid tablets are all forms of moving your designs skyward. Just about anything that generates gas can be used. Hmm, that reminds me of a joke about Uranus.

Finally, if you can ever locate a used copy of Brinley's *Rocket Manual for Amateurs*, buy it. The design theory and Remington's illustrations will be worth the price. Whatever it is.

Project 11. A Plan for a Rocket Fighter

What You Need:

- Balsa
- Tissue
- Foil
- Metal clip
- Glue
- Paint

Resources:

- Jetex Organization—jetex.org
- Rapier Pardubice—www.rapier.cz/index.htm

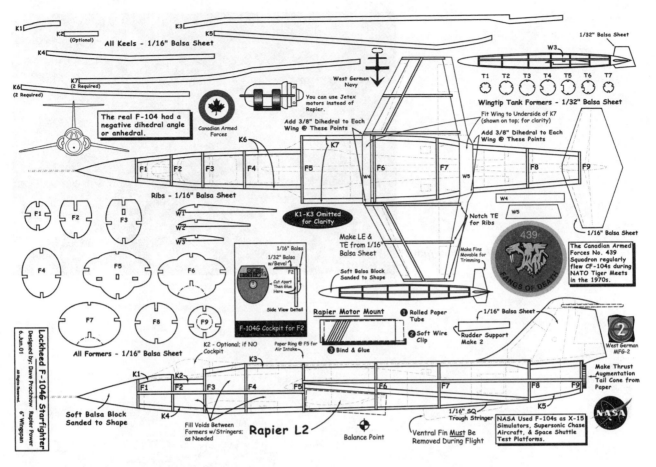

K1

K2 (Optional)

K3

K5

All Keels - 1/16" Balsa Sheet

K4

1/32" Balsa Sheet

W3

T1 T2 T3 T4 T5 T6 T7

K6 (2 Required)

K7 (2 Required)

Wingtip Tank Formers - 1/32" Balsa Sheet

The real F-104 had a negative dihedral angle or anhedral.

West German Navy

Canadian Armed Forces

You can use Jetex motors instead of Rapier.

Add 3/8" Dihedral to Each Wing @ These Points

Fit Wing to Underside of K7 (shown on top; for clarity)

Add 3/8" Dihedral to Each Wing @ These Points

F1 F2 F3 F4 K6 K7 F5 W4 F6 F7 W5 F8 F9

Ribs - 1/16" Balsa Sheet

W4

W5

1/16" Balsa Sheet

F1 F2 F3

W1'
W2'
W3'

K1-K3 Omitted for Clarity

Make LE & TE from 1/16" Balsa Sheet

Notch TE for Ribs

Make Fins Movable for Trimming

The Canadian Armed Forces No. 439 Squadron regularly flew CF-104s during NATO Tiger Meets in the 1970s.

439 FANGS OF DEATH

F4 F5 F6

1/16" Balsa

1/32" Balsa w/Bevel

Cut Apart Then Glue Here

F2

Side View Detail

Soft Balsa Block Sanded to Shape

F7 F8 F9

F-104G Cockpit for F2

All Formers - 1/16" Balsa Sheet

K2 - Optional; if NO Cockpit

Rapier Motor Mount
❶ Rolled Paper Tube
❷ Soft Wire Clip
❸ Bind & Glue

Paper Ring @ F5 for Air Intake

1/16" Balsa Sheet

Rudder Support Make 2

West German MFG-2

Lockheed F-104G Starfighter

Designed by: Dave Prochnow
6.Jun.01
All Rights Reserved.

Rapier Power
6" Wingspan

K1
K2
K3

F1 F2 F3 F4 F5 F6 F7 F8 F9

Make Thrust Augmentation Tail Cone from Paper

K5

Soft Balsa Block Sanded to Shape

K4

Fill Voids Between Formers w/Stringers; as Needed

Rapier L2

Balance Point

1/16" SQ Trough Stringer

Ventral Fin **Must** Be Removed During Flight

NASA Used F-104s as X-15 Simulators, Supersonic Chase Aircraft, & Space Shuttle Test Platforms.

NASA

Figure 11-1   *A plan for building a rocket-powered Lockheed F-104G Starfighter.*

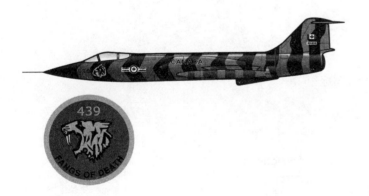

439 FANGS OF DEATH

Figure 11-2   *A great orange/black paint scheme for your rocket plane model.*

Figure 11–3 *This completed model flew very well and very fast.*

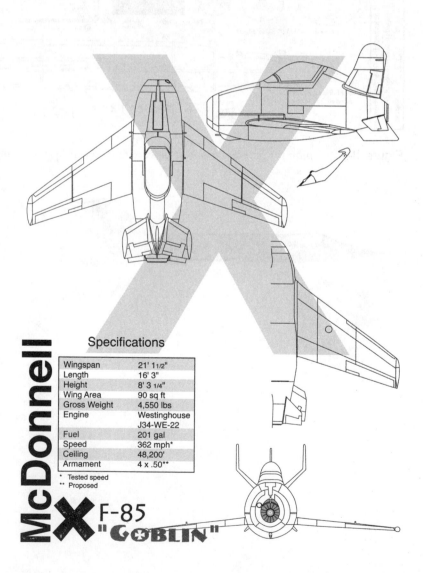

Specifications

Wingspan	21' 1 1/2"
Length	16' 3"
Height	8' 3 1/4"
Wing Area	90 sq ft
Gross Weight	4,550 lbs
Engine	Westinghouse J34-WE-22
Fuel	201 gal
Speed	362 mph*
Ceiling	48,200'
Armament	4 x .50**

* Tested speed
** Proposed

McDonnell XF-85 "Goblin"

Figure 11–4 *The XF-85 is another excellent rocket-powered model candidate.*

Project 12. How to Design Your Own Rocket

What You Need:

- Balsa
- Foam balls
- Paper tubes
- Parachute
- Rubber band

Resources:

- Apogee Rockets—www.apogeerockets.com
- Estes Rockets—www.estesrockets.com
- National Association of Rocketry—www.nar.org
- Quest Model Rocketry—www.questaerospace.com

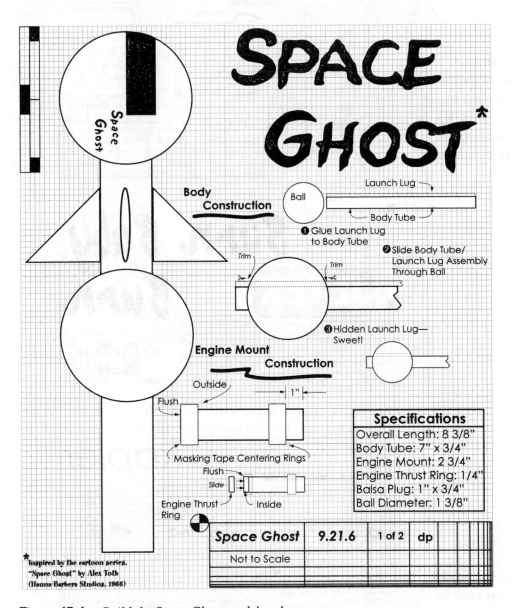

Figure 12–1 *Build the Space Ghost model rocket.*

Figure 12–2 *Watch out for four construction points: (1) make sure the fins are properly aligned, (2) securely install the shock cord, (3) run the launch lug through both Styrofoam® spheres, and, finally, (4) pick the perfect propulsion system.*

SPACE GHOST

Launch Lug

Balsa Plug

Screw Eye

Fin Attachment

Shock Cord Installation

90°

Make 2 Slits in Body Tube

Insert Shock Cord & Glue

Engine Thrust Ring

Engine Mount

Full-Size Fin Template

Launch Lug

Sanding Profile

Apply Gloss White Paint for Finish

Grain

Use 1/8" Thick Balsa

Specifications	
Overall Length: 8 3/8"	
Body Tube: 7" x 3/4"	
Engine Mount: 2 3/4"	
Engine Thrust Ring: 1/4"	
Balsa Plug: 1" x 3/4"	
Ball Diameter: 1 3/8"	

Space Ghost	9.21.6	2 of 2	dp
Not to Scale			

Use B6-4 Engines for Flights of 500-800 ft.

Figure 12–3 *Several exciting model rocket propulsion systems are ideal for experimenting with in your designs.*

Burn Baby, Burn

JETEX

Jetex was patented in the United Kingdom in March 1948 by Wilmot, Mansour & Co., Ltd. This propulsion system used an aluminum alloy motor equipped with a safety mechanism and fueled with solid pellets. The first Jetex motors went on sale in June 1948.

In March 1950, American Telasco became the official US distributor for Jetex. During the height of production, eight motors were produced: Atom 35, 50 & 50B/C, 100, Jetmaster, 150, 200, 350, and Scorpion 600. Each model roughly equalled the rated thrust in ounces. For example, the Jetex 50 = .5 oz. of thrust.

D. Sebel and Company purchased the Jetex model engine line in 1956. These new motors featured steel cases. Similarly, Sebel began manufacturing their own line of hotter, more powerful Jetex fuel in 1959.

In 1986, Powermax launched a new line of solid propellant model engines called Jet-X. Megamodels of the United Kingdom sells Jet-X motors and supplies (www.megamodels.co.uk).

Rapier motors are available in three power ratings: L1, L2 and L-2HP. Rapier motors might not be powerful enough to launch a rocket vertically into space.

Rapier

Dr. Jan Zigmund of the Czech Republic created Rapier motors. Rapier motors are built from a cardboard tube that is filled with a solid propellant fuel, and plugged with a ceramic end cap. Shorty's Basement of Ohio sells Rapier motors (www.shortysbasement.com).

ESTES

1/2A6-2

ESTES

In 1958, Vernon Estes designed the first commercial model rocket engine. Since the 1960s, Estes Industries has been a leader in model rocketry. Currently, there are over 25 different model rocket engines manufactured and sold by Estes.
www.estesrockets.com

Never launch a model rocket that weighs more than one pound or contains more than 4 ounces of propellant. If you do, you will get a knock at your door from a whole slew of "G-Men."

Propulsion Systems	9.20.6	1 of 1	dp
	Rev.		

Figure 12–4 *The materials needed for building Space Ghost.*

Figure 12–5 *Find the center for each Styrofoam sphere.*

Figure 12–6 *Prepare your motor mount.*

Figure 12–7 *Carefully drill a hole through the center of the first sphere.*

Figure 12–8 *Don't drill a hole all the way through the second sphere.*

Figure 12–9 *Assemble the model and try a couple of test flights.*

Project 13. Plop, Plop, Fizz, Fizz, ZOOM

What You Need:

- 35mm film canister
- Antacid tablet

Resources:

- The Space Place—spaceplace.jpl.nasa.gov/en/kids/

Figure 13–1 *Antacid tablets like Alka-Seltzer® are potent rocket-propellant systems (I kid you not).*

Figure 13–2 *In addition to the antacid tablets, you also need an empty 35mm film canister. These two items are your motor mount and propellant. You also need a rocket body tube that can snuggly and safely hold the motor mount (film canister).*

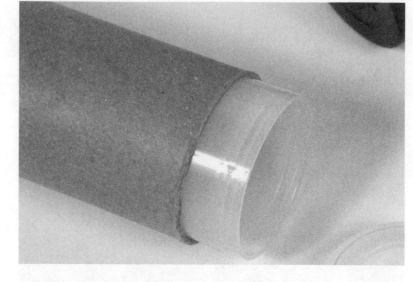

Figure 13–3 *In my tests, I found that Fujifilm canisters (Fuji Photo Film Co., Ltd.) are the best motor mounts.*

Figure 13–4 *Conversely, the film canisters manufactured by Eastman Kodak® don't work well as motor mounts.*

Figure 13–5 *Don't believe me; test various film canisters for yourself. Just watch out. The rocket propulsion system is easily capable of launching a lightweight model 10 to 20 feet into the air.*

Project 14. Build a Hydrogen-Oxide Rocket

What You Need:

- Plastic water bottle
- PVC tubing
- Bicycle pump

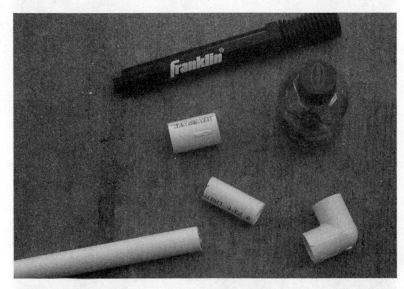

Figure 14–1 *You can build a water-powered rocket from an old water bottle, some PVC plumbing tubing, a rubber seal, fasteners, and a bicycle pump.*

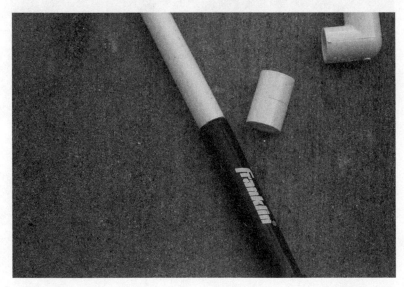

Figure 14–2 *The pump is connected to the "launch pad" via PVC tubing.*

Figure 14–3 *This "launch pad" holds the water-bottle rocket while air is pumped into the rocket.*

Figure 14–4 *Pump 'er up and pull the retaining fastener that holds the rocket on the launch pad. Blast off!*

Project 15. Put Your Own Eye in the Sky

What You Need:

- Model rocket
- Small digital camera

Figure 15–1 *A CCD (Charge-Coupled Device) sensor from a discarded 3Com Webcam.*

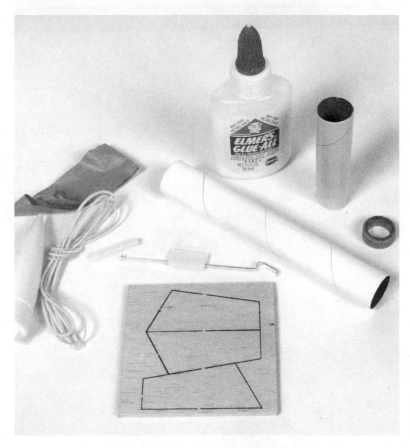

Figure 15–2 *The materials needed for building a simple CCD-equipped model rocket (based on a Quest™ Aerospace (www.questaerospace.com) model rocket kit).*

Figure 15-3 *Connect the ends of your elastic shock cord.*

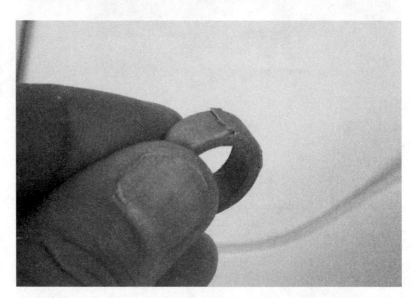

Figure 15-4 *Cut a small groove in the motor mount block.*

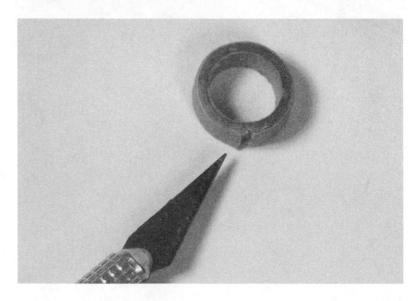

Figure 15-5 *Slowly work the groove into a V-shaped slot.*

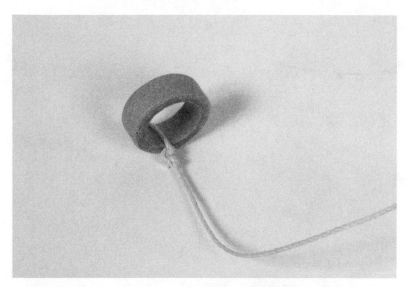

Figure 15–6 *Attach the shock cord to the motor mount block.*

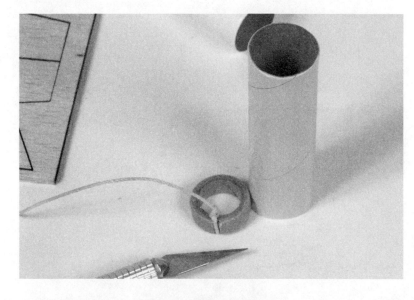

Figure 15–7 *Measure the depth of the motor mount block on the motor mount.*

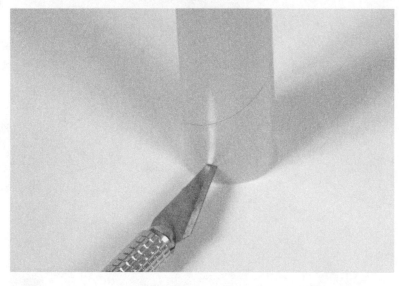

Figure 15–8 *Cut a small slit in the motor mount.*

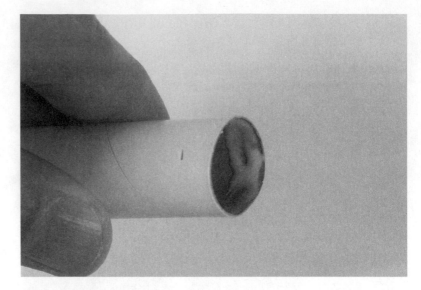

Figure 15–9 *Apply some glue inside the motor mount on the end with the slit.*

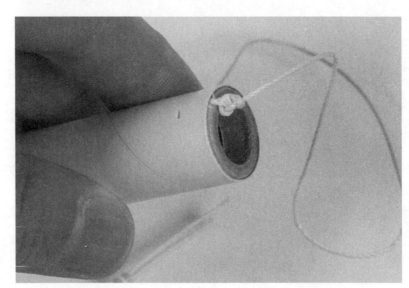

Figure 15–10 *Slip the motor mount block into the glued end.*

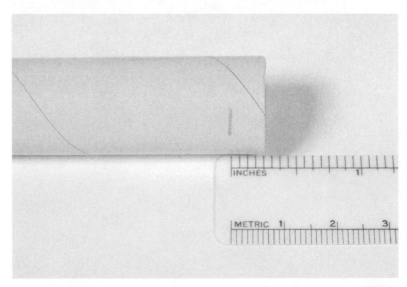

Figure 15–11 *Measure the motor mount installation depth.*

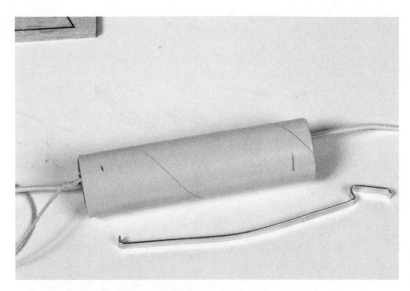

Figure 15–12 *Slightly bend the motor retaining clip.*

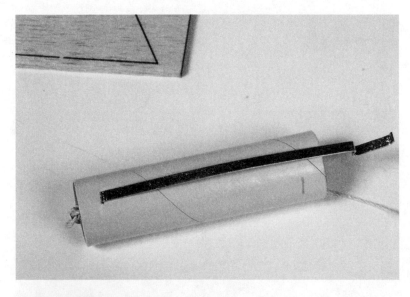

Figure 15–13 *Loosely install the motor retaining clip.*

Figure 15–14 *Apply some glue inside one end of the main rocket body tube.*

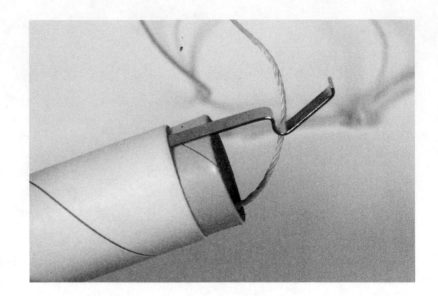

Figure 15–15 *Slide the motor mount into the glued end of the body tube.*

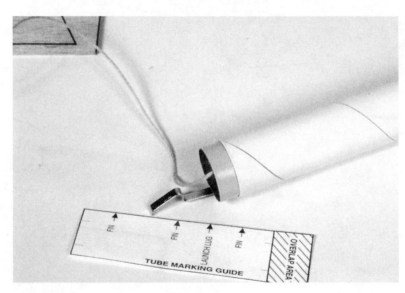

Figure 15–16 *Determine the location for the fins and launch lug.*

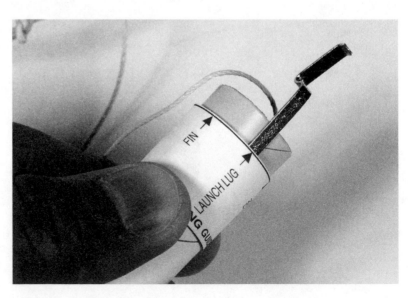

Figure 15–17 *Line up the marking guide with the motor retaining clip, matching the mark for the launch lug.*

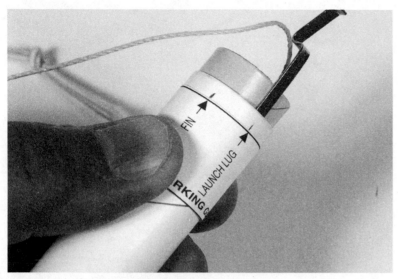

Figure 15–18 *Mark the main rocket body tube for the fins and launch lug.*

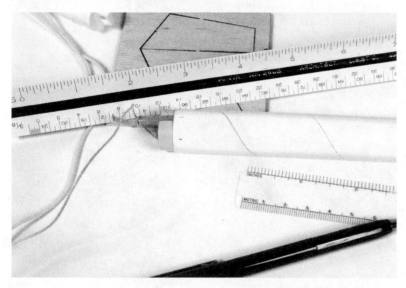

Figure 15–19 *Use a grooved "architect's scale" for extending the fin and launch lug marks along the length of the main rocket body tube.*

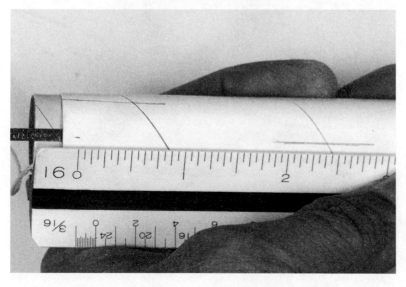

Figure 15–20 *Cradle the body tube inside the groove of the scale, and extend the fin and launch lug marks.*

Figure 15–21 *Cut out, trim, and sand three fins to a smooth finish.*

Figure 15–22 *Glue the first fin to the main model rocket body tube along the previously marked line.*

Figure 15–23 *Pull the shock cord up and through the "other" end of the body tube.*

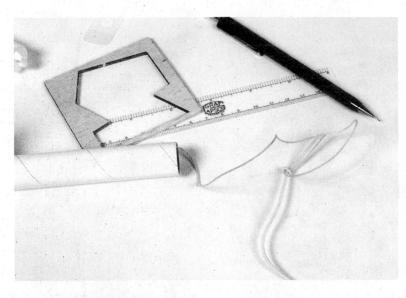

Figure 15–24 *Carefully pull the entire shock cord up and out of the body tube. Don't pull so hard that you dislodge the motor mount block.*

Figure 15–25 *Continue to glue the remaining two fins on to the body tube.*

Figure 15–26 *Add glue fillets to both sides of each fin. This action reinforces each fin, as well as streamlines the air flow along the rocket's body.*

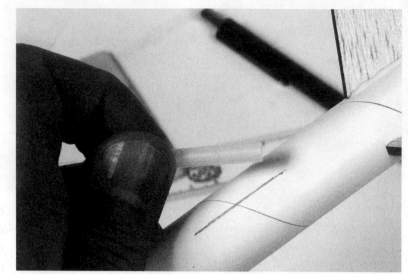

Figure 15–27   *Attach the launch lug to the main rocket body tube.*

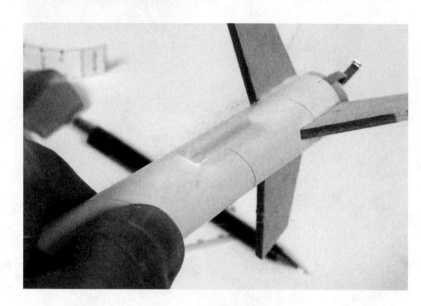

Figure 15–28   *Make sure the launch lug is properly aligned with the motor-retaining clip.*

Figure 15–29   *Prepare the rocket's recovery system.*

Figure 15–30 *A plastic gripper tab will be used for attaching the streamer to the rocket's shock cord.*

Figure 15–31 *Attach the gripper tab to the streamer.*

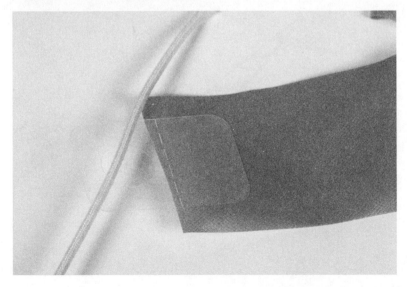

Figure 15–32 *Run the shock cord through the hole in the gripper tab.*

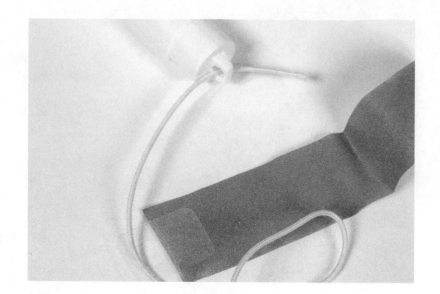

Figure 15–33 *Run the shock cord through the nose cone.*

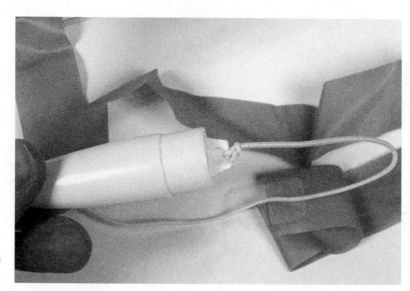

Figure 15–34 *Tie the shock cord to the nose cone.*

Figure 15–35 *Finish and paint your completed model rocket.*

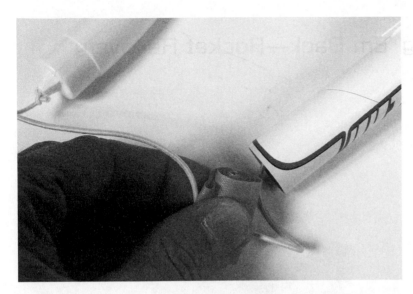

Figure 15–36 *Install the recovery system.*

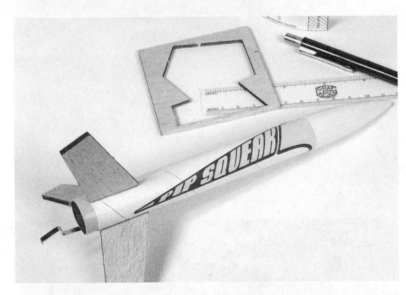

Figure 15–37 *Ready for launch.*

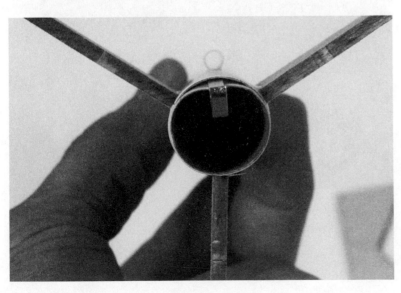

Figure 15–38 *Check fin alignment prior to every launch.*

Project 16. Bring 'Em Back—Rocket Recovery

What You Need:

- Model rocket
- Streamer
- Parachute

Figure 16–1 *The materials needed for a parachute recovery system.*

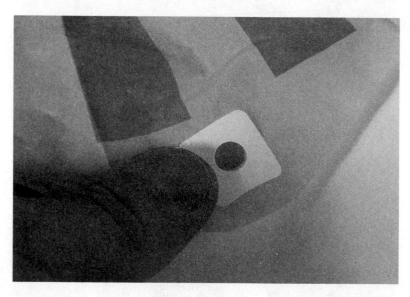

Figure 16–2 *Use plastic reinforcement rings for preparing the plastic "chute."*

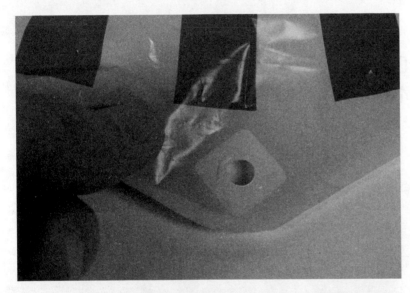

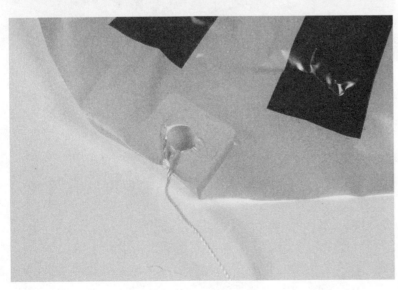

Figure 16–3 *Spread the plastic out and attach all the reinforcement rings.*

Figure 16–4 *Make sure no bubbles form in the plastic parachute around each reinforcement ring.*

Figure 16–5 *Tightly bind each end of the shroud line to the parachute.*

Figure 16–6 *Gather all the shroud lines and tie the completed parachute assembly to the rocket's nose cone.*

Project 17. Get Higher with Boosters and Multistaged Rockets

What You Need:

- Model rocket
- Additional paper tubes

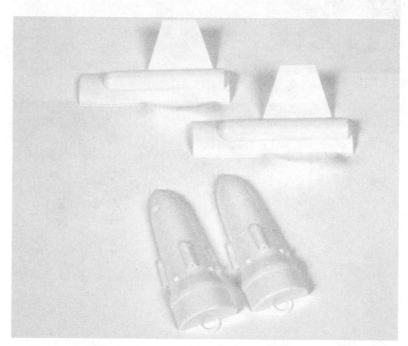

Figure 17–1 *You can add realism to your rocket with spare parts left over from conventional plastic model kits.*

Figure 17-2 *Keep these plastic parts away from the model rocket motor's hot exhaust gases.*

Figure 17-3 *Multiengine rockets can dramatically increase the performance of your designs.*

What You Need:

- Model rocket motor test stand
- Cardboard tubes
- Cardboard

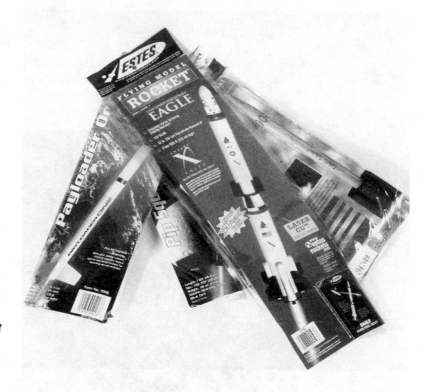

Figure 18–1 *Commercial model rocket kits are heavily tested prior to their sale to the public.*

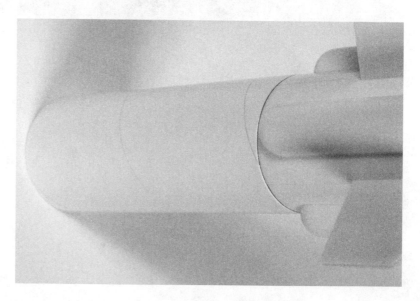

Figure 18–2 *Build your own model rocket test stand to evaluate your designs.*

Project 19. A Plan for a Scale Rocket Glider with Remote Control

What You Need:

- Balsa
- Foam slab
- Glue
- Paint
- Two-channel radio control system
- Two micro-sized servos

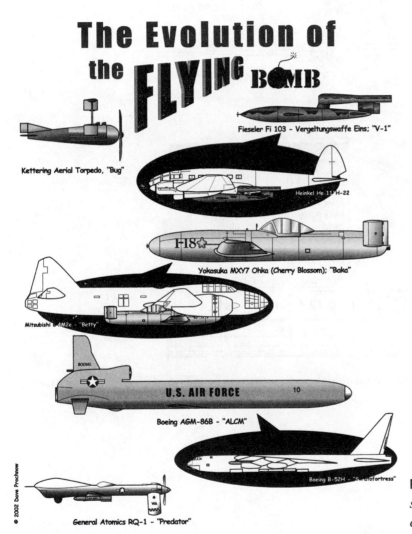

The Evolution of the FLYING BOMB

Kettering Aerial Torpedo, "Bug"

Fieseler Fi 103 - Vergeltungswaffe Eins; "V-1"

Heinkel He 111/H-22

Yokosuka MXY7 Ohka (Cherry Blossom); "Baka"

Mitsubishi G4M2e - "Betty"

Boeing AGM-86B - "ALCM"

Boeing B-52H - "Stratofortress"

General Atomics RQ-1 - "Predator"

© 2002 Dave Prochnow

Figure 19–1 *Here are some great subjects for designing your own remote-controlled model rocket.*

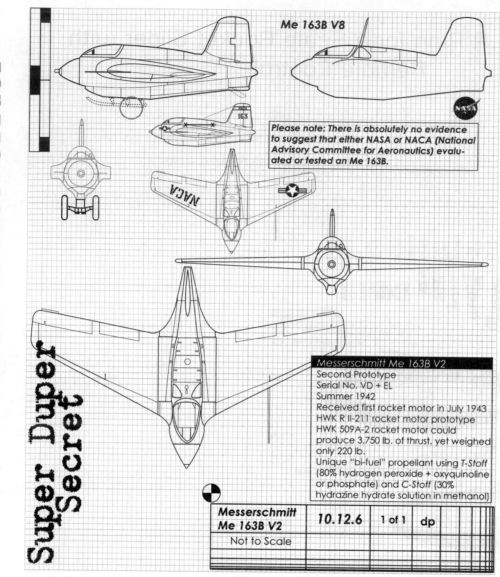

Me 163B V8

Please note: There is absolutely no evidence to suggest that either NASA or NACA (National Advisory Committee for Aeronautics) evaluated or tested an Me 163B.

Super Duper Secret

Messerschmitt Me 163B V2
Second Prototype
Serial No. VD + EL
Summer 1942
Received first rocket motor in July 1943
HWK R II-211 rocket motor prototype
HWK 509A-2 rocket motor could produce 3,750 lb. of thrust, yet weighed only 220 lb.
Unique "bi-fuel" propellant using *T-Stoff* (80% hydrogen peroxide + oxyquinoline or phosphate) and *C-Stoff* (30% hydrazine hydrate solution in methanol)

Messerschmitt Me 163B V2	10.12.6	1 of 1	dp		
Not to Scale					

Figure 19–2   *The Messerschmitt Me 163B is a great high-performance, rocket-powered glider for using as the basis of a remote-controlled model.*

What You Need:

- Model rocket with payload carrier
- Insect victims, err, test subjects

Figure 20–1 *Only two extra parts are needed for building a payload-lifting model rocket—a connector or joiner tube and a payload tube.*

Figure 20–2 *The nose cone connects to the payload tube without the typical recovery system attachment. Generally, the nose cone can be glued in place inside one end of the payload tube.*

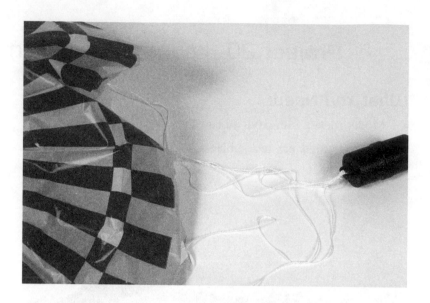

Figure 20–3 *The recovery system is attached to one end of the connector or joiner tube. Unless you are a bigger evil genius than even Penelope, you should strive for a parachute recovery system.*

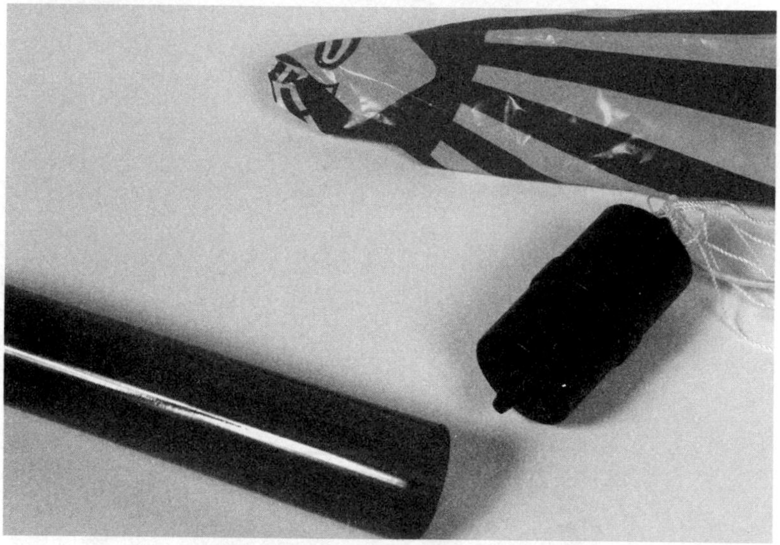

Figure 20–4 *Add your payload and slip the connector tube in place on the opposite end of the payload tube.*

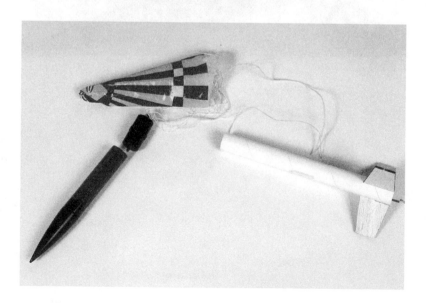

Figure 20–5 *The other end of the shock cord should be attached to the model rocket tube.*

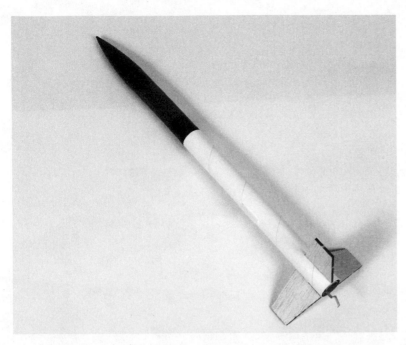

Figure 20–6 *Ready for liftoff.*

Figure 20–7 *Bring 'em back alive. Crickets and grasshoppers make ideal "astronauts" for payload experiments.*

Chapter Three

Take a Left at Alpha Centauri

Look up; look around yourself. No matter where you call home on this big sphere of ours, you should be able to see one or two hundred stars every clear night. But how many stars can you actually see?

Let's find out how many stars are in *your* night's sky with the "bonus" star-counting project. Based on the popular NASA Star Count activity, this project will give you a rough, average number of stars you can see at night right where you live.

What You Need:

- Paper tube
- Tape
- Calculator

Resources:

- AIRNow—www.airnow.gov/index.cfm? action= airnow.national
- National Aeronautics and Space Administration—www.nasa.gov
- NASA Star Count, Student Observation Network—www.nasa.gov/audience/ foreducators/ starcount/home/index.html

Figure 3–1   *The essential star-gazing field kit (from left to right): Great Red Spot Junior Red LED Shake Flashlight, Magellan eXplorist 210, and Night Owl Optics Night Vision scope.*

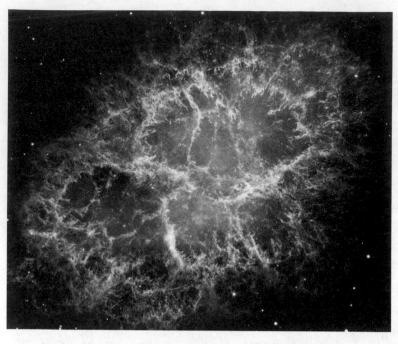

Figure 3–2   *This Hubble Space Telescope WFPC2 image—among the largest ever produced by the Earth-orbiting observatory—shows the most detailed view so far of the entire Crab Nebula. The Crab is arguably the single most interesting object, as well as one of the most studied, in all of astronomy. The Hubble image of the Crab was assembled from 24 individual exposures taken with the NASA/ESA Hubble Space Telescope (image courtesy of NASA/ESA/JPL/ Arizona State University).*

Project 21. Night of the Trifids

What You Need:

- Nikon FM10
- Vivitar 500mm lens

Resources:

- Nikon, Inc.—www.nikonusa.com

Figure 21–1   *These two cases contain the essential elements needed for deep-space photography.*

Figure 21–2 *Vivitar 500mm f/8 mirror lens.*

Figure 21–3 *Filters for a mirror lens are attached to the rear lens element.*

Figure 21–4 *This mirror lens uses a "T-Mount" for attaching it to a camera. The interchangeable T-Mount (for example, just screw a different mount onto the lens; these mounts are available from a wide variety of different camera manufacturers) enables this lens to be used with different camera bodies.*

Figure 21–5 *A Nikon T-Mount enables this Vivitar 500mm mirror lens to be attached to this Nikon FM10 camera.*

Figure 21–6 *Attach the camera and mirror lens to a stable tripod for taking some deep-space photographs. This photograph was taken during the daylight hours.*

What You Need:

• Starry Night Pro

Resources:

• Imaginova®—www.imaginova.com
• Starry Night Store—www.starrynight.com

Figure 22–1 *Starry Night Pro by Imaginova Corporation.*

Figure 22–2 *Starry Night Pro is the perfect star-gazing companion. New features in this software enable you to wirelessly control supported telescopes through Bluetooth® technology.*

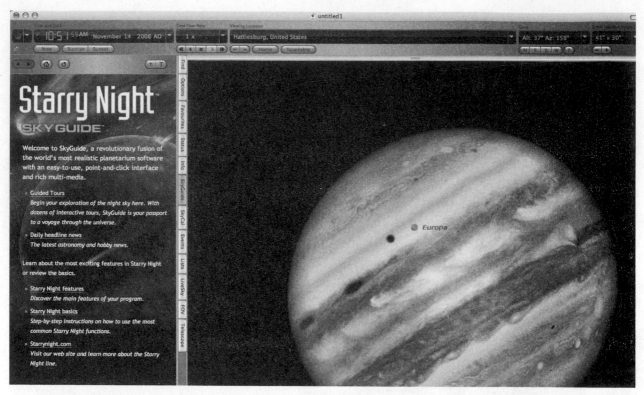

Figure 22-3   *Special events like this Europa transit of Jupiter are available in Starry Night Pro.*

Figure 22-4   *A toolbar showing time, date, time flow, viewing location, gaze, and fields of view runs across the top of your computer's screen.*

Figure 22-5 *Once you properly configure Starry Night Pro for your home viewing location, the main screen shows you what you could be seeing outside your living room window. That is, if you'd get your nose out of the computer's screen.*

Figure 22-6 *Increasing the time flow can quickly change your sky view to a specific time or event. Just toggle the play/stop buttons to control all the heavenly action.*

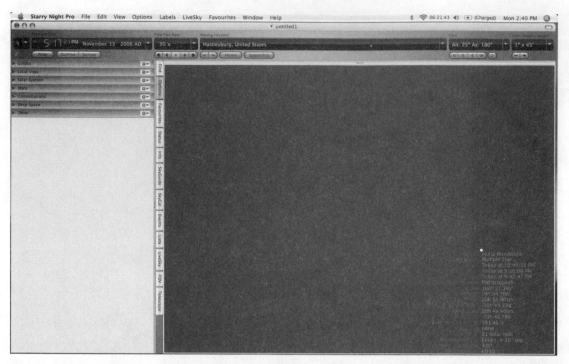

Figure 22–7 *Hold your cursor over any celestial object to receive complete and thorough information about any satellite, star, constellation, planet, moon, and comet. And, if the extensive Starry Night Pro database doesn't contain your selected celestial object, you can add it to your own database.*

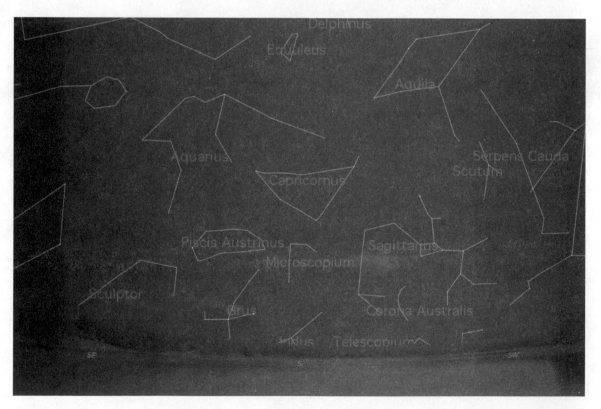

Figure 22–8 *You can quickly locate any constellation, complete with names and stick figure outlines.*

Figure 22-9 *You can also opt for displaying classical constellation illustrations for your night sky view. Or, get really creative and draw your own constellation stick figure art. Now you can make Orion really look like, well, Orion.*

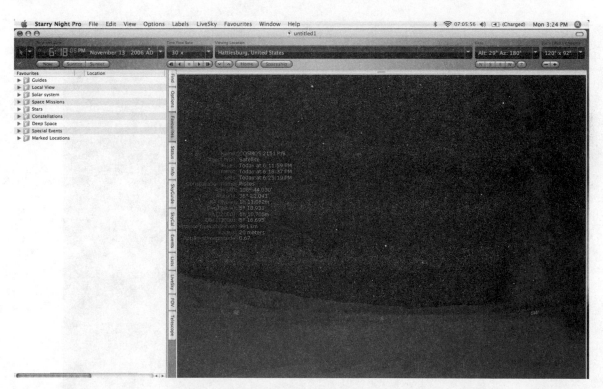

Figure 22-10 *Trying to track down something you just saw (or, thought you saw) zip across your field-of-view is a snap with the contextual pop-up information tips.*

Figure 22–11 *Graph the magnitude, altitude, separation, distance, elongation, and angular size of various objects, and then compare them to each other. Shown here are the plots for the Sun and the Moon.*

Figure 22–12 *If you have a hankering to find a specific object, search the Starry Night Pro databases.*

Name			Alt	Kind
Sun		ⓘ	–35°	Sun
Mercury	☐	ⓘ	–42°	Planet
Venus	☐	ⓘ	–32°	Planet
▼ Earth	☐	ⓘ		Planet
The Moon	☐	ⓘ	–47°	Moon
▶ Satellites				
▶ Mars	☐	ⓘ	–40°	Planet
▶ Jupiter	☐	ⓘ	–30°	Planet
▶ Saturn	☐	ⓘ	–39°	Planet
▶ Uranus	☐	ⓘ	49°	Planet
▶ Neptune	☐	ⓘ	34°	Planet
▶ Pluto	☐	ⓘ	–4°	Planet
▶ Space Missions				
▶ Asteroids				
▶ Comets				

Figure 22–13 *A selected object is displayed on the main screen.*

Project 23. Build a Star Catalog

What You Need:

- Starry Night Pro

Resources:

- Space.com—www.space.com

In Starry Night Pro, your star catalog will be known as a database. This database is created with a text editor. Once this database file is written, Starry Night Pro will import, convert, and install the database. After restarting the application, your new star catalog database will be ready to use.

Project 24. Explore Deep Space

What You Need:

- Starry Night Pro Plus

Resources:

- LiveScience.com—www.livescience.com

Inside Starry Night Pro Plus are several exclusive features that are ideally suited to exploring deep space: AllSky CCD Mosaic, AllSky options, and nebula outlines.

Project 25. Take a 3-D Adventure into Space

What You Need:

- Uncle Milton 3-D Adventure Projector
- Uncle Milton "Journey into Space" Adventure Pack

Resources:

- Uncle Milton Toys—www.unclemilton.com

Figure 25–1 *Uncle Milton 3-D Adventure Projector.*

Figure 25–2 *You can project 3-D images on any flat, light-colored surface with the Adventure Projector.*

Figure 25–3 *Avoid poor image quality by keeping the projector parallel with the viewing surface.*

Figure 25–4 *Turn the built-in lens to focus the 3-D images.*

Figure 25–5 *3-D glasses are supplied with the 3-D Adventure Projector. You must wear these glasses to see the 3-D effect.*

Chapter Four

Make Contact

Who isn't amazed with the tones generated when you tune a radio? By turning a tuning wheel (or pushing a SEEK/SCAN button on digital radios), a delightful aural smorgasbord of static, talk, and music greets your ear. Even more miraculous, all this magic is from waves of electromagnetic energy that you can't even see ... or feel.

Now imagine your excitement when the sounds you hear on your radio are actually the sounds of space (see Figure 4-1). These sounds are the domain of radio astronomy (see Figure 4-2).

Before you take a deep space plunge into radio astronomy, however, maybe you should test your radio skills by receiving a common radio signal that emanates from the central U.S.

A special radio station (WWVB) in Fort Collins, Colorado, maintained by the Time and Frequency Division of the National Institute of Standards and Technology (NIST), allows for the synchronization of time around the country.

Pumping out a 50 kW signal on a frequency of 60 kHz, WWVB is able to transmit a time synchronization standard to radio-controlled clocks (RCC) with a stated accuracy of within 1 second or less. Even better, once an RCC is turned on, the user can just walk away and forget it. No time setting is required ... ever.

One of these complete time-code cycles consists of an on-time marker (OTM), which is sent every second by lowering the 60 kHz frequency by 10dB in synchronization with a Coordinated Universal Time (UTC) second. The bit truth condition is identified by the duration of the low-power OTM. For example, a 0 bit is held low for 200 ms, whereas a 1 bit is held low for 500 ms. Frame markers are then sent every 10 s and held for 800 ms.

Finally, did you know that "leap seconds" are periodically added to the WWVB transmission? These insertions typically occur on June 30

Figure 4-1 *Aerial view of the Very Large Array (VLA), looking north-northeast; the antennas are in their closest configuration (D configuration). The National Radio Astronomy Observatory (NRAO) is a facility of the National Science Foundation, operated under cooperative agreement by Associated Universities, Inc. (AUI). (Photo by Dave Finley: Courtesy NRAO/AUI.)*

Figure 4-2 *This true-color simulated view of Jupiter is composed of four images taken by NASA's Cassini spacecraft on December 7, 2000. To illustrate what Jupiter would have looked like if the cameras had a field-of-view large enough to capture the entire planet, the cylindrical map was projected onto a globe. The resolution is about 144 kilometers (89 miles) per pixel. Jupiter's moon Europa is casting the shadow on the planet.*

Cassini is a cooperative mission of NASA, the European Space Agency, and the Italian Space Agency. JPL, a division of the California Institute of Technology in Pasadena, manages Cassini for NASA's Office of Space Science, Washington, D.C. (image courtesy of NASA/JPL/University of Arizona).

and/or December 31. This added second makes the final minute of the day 61 seconds long. For a good RCC to properly show a leap second, the number 60 should be able to be displayed in the seconds' field of the clock. A lower-priced RCC will probably just hold the display of 59 seconds for two seconds. Now, exactly what you do with the extra second is up to you. This same principle holds for the display of leap years. In this case,

however, the RCC date display must properly indicate a leap year date as February 29.

Now get out there and QSL (radio contact confirmation) WWVB in Fort Collins.

Resources:

- National Institute of Standards and Technology—tf.nist.gov/general/publications.htm

Project 26. Build a Radio That Doesn't Require Power

What You Need:

- Diode
- Tuning capacitor
- Wire, lots of wire

Figure 26–1   *A crystal radio kit.*

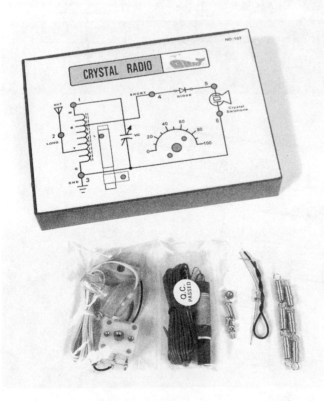

Figure 26–2   *Do you notice something missing from this kit? No battery or other power supply.*

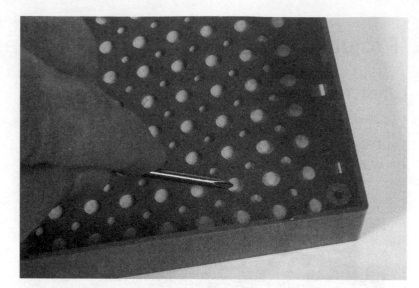

Figure 26–3 *Punch holes in the front paper face for holding the components.*

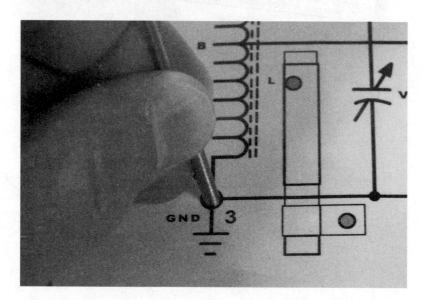

Figure 26–4 *A small screwdriver is perfect for making smooth holes.*

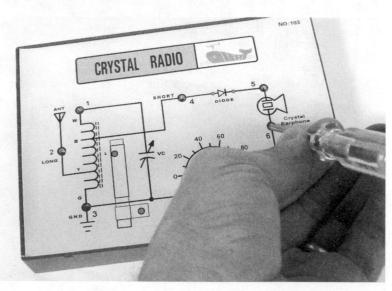

Figure 26–5 *Each of the large holes is numbered.*

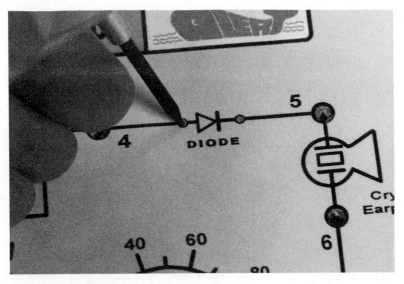

Figure 26-6 *Use an awl for making the smaller holes.*

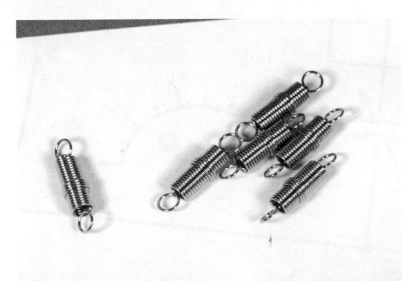

Figure 26-7 *No soldering required. These spring clips are used for joining the components together.*

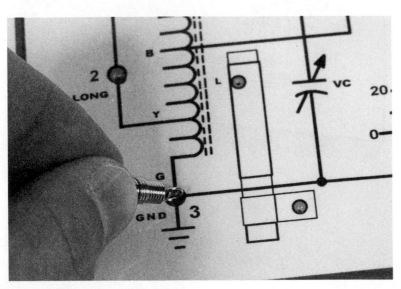

Figure 26-8 *Push one spring clip into each of the large numbered holes.*

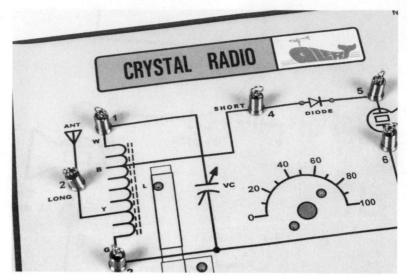

Figure 26–9 *Each spring clip is inserted into each hole until it hits its stop collar.*

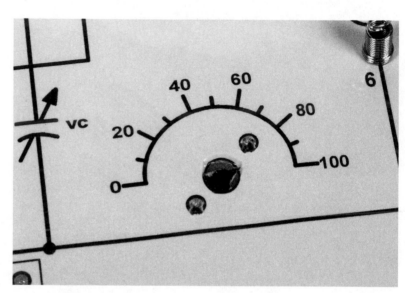

Figure 26–10 *Make three holes for the tuning capacitor.*

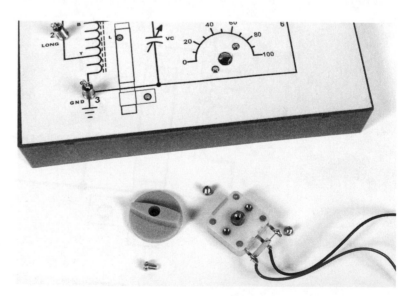

Figure 26–11 *The capacitor will sit underneath the kit's case, while the tuning knob will ride on top of the front paper face.*

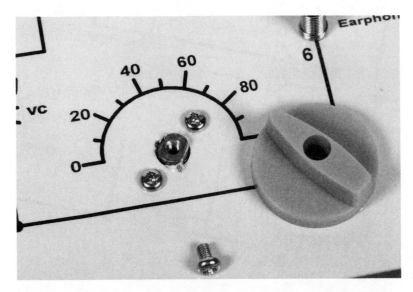

Figure 26–12   *Two screws hold the tuning capacitor.*

Figure 26–13   *The knob is attached to the capacitor.*

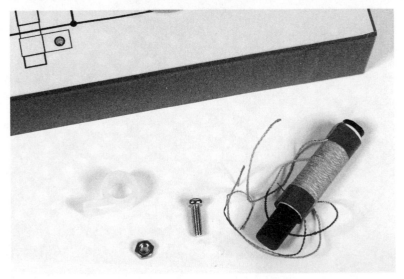

Figure 26–14   *The ferrite core coil.*

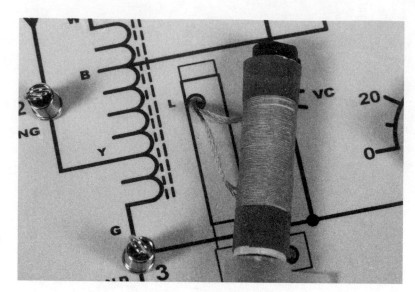

Figure 26–15   *Feed the four coil leads through the small L hole on the front paper face.*

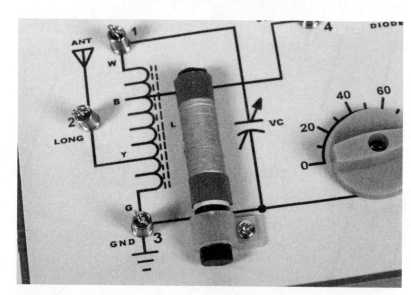

Figure 26–16   *Use the cable hold-down clamp for attaching the coil to the front paper face.*

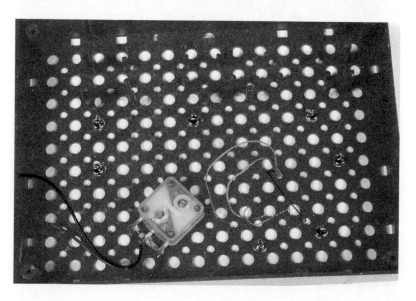

Figure 26–17   *The underside of the crystal radio-kit case.*

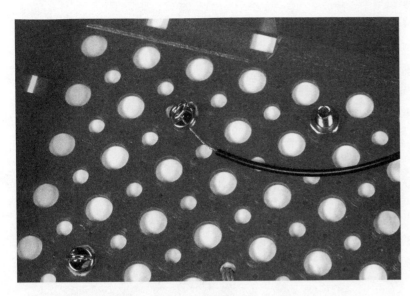

Figure 26–18 *The spring clips are used for wiring the components together.*

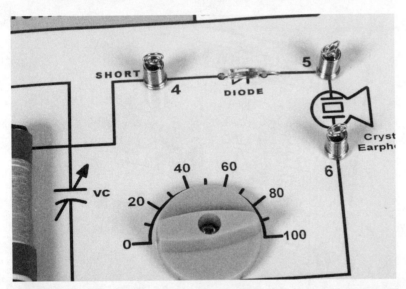

Figure 26–19 *Install the diode.*

Figure 26–20 *The crystal radio is completely wired.*

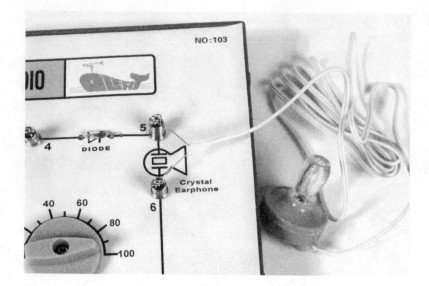

Figure 26–21 *Attach the earphone.*

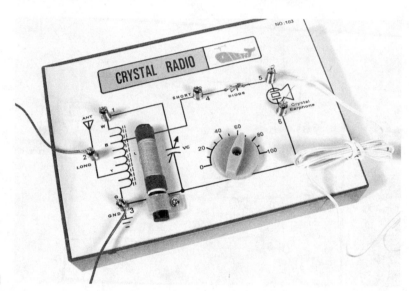

Figure 26–22 *Connect an antenna and ground wire, and then the crystal radio is ready for operation.*

Project 27. Use a Shortwave Radio

What You Need:

- Sangean ATS505P

Resources:

- C. Crane Company, Inc.—www.ccrane.com

A low-cost shortwave radio can be an exciting door to a whole new frontier—long-distance listening.

Project 28. Add an Antenna

What You Need:

- 10 feet of wire

One of the best places in your home for installing your new antenna is in your attic space. Just clip the antenna along the bottom of several roof rafters. Make sure that you use insulated clips.

Project 29. Build an Antenna for Star Listening

What You Need:

- 30–40 meters of wire

Figure 29–1 *In a pinch, you can use trees as your poles for stringing your antenna.*

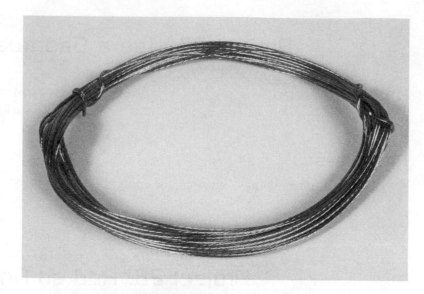

Figure 29-2 *Bare, stranded copper wire makes a great antenna.*

Figure 29-3 *Keep the wire strands twisted together when stringing your antenna.*

Project 30. By Jove, Now You've Got It

What You Need:

- NASA JOVE radio receiver

Resources:

- NASA Radio JOVE Project—
radiojove.gsfc.nasa.gov

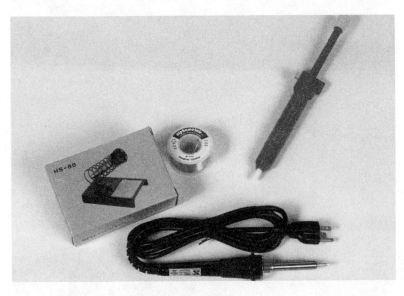

Figure 30–1 *You need some basic soldering equipment for building the Radio JOVE receiver.*

Figure 30–2 *One of the best electronic kits you will ever build ... and at a terrific price: the Radio JOVE Receiver/Antenna kit.*

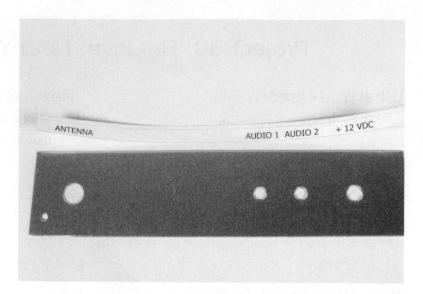

Figure 30–3 *Prepare the chassis rear panel.*

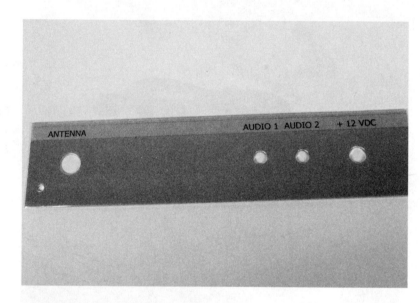

Figure 30–4 *Attach the dry transfer decal to the chassis rear panel.*

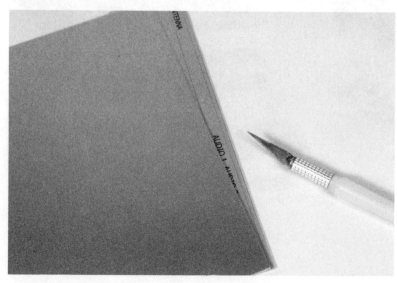

Figure 30–5 *Trim the decal using the edge of the chassis bottom panel.*

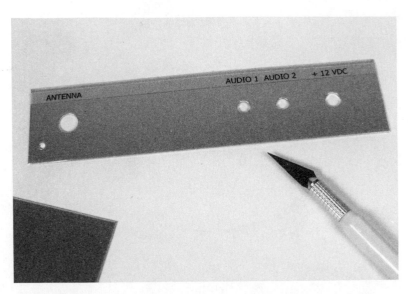

Figure 30–6 *Set the rear panel aside.*

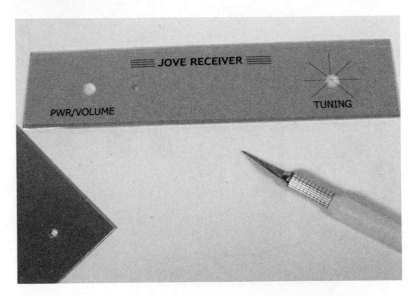

Figure 30–7 *Attach the dry transfer decal to the chassis front panel.*

Figure 30–8 *Cut small openings in the decal for the Pwr/Volume switch, LED, and Tuning potentiometer. Set the front panel aside.*

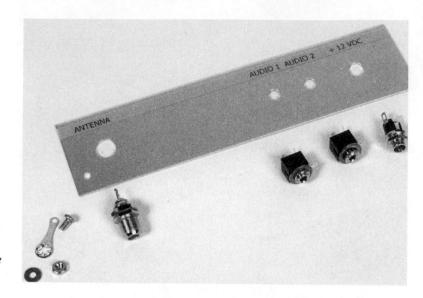

Figure 30–9 *Return to the chassis rear panel, and then locate the hardware jacks and grounding lug.*

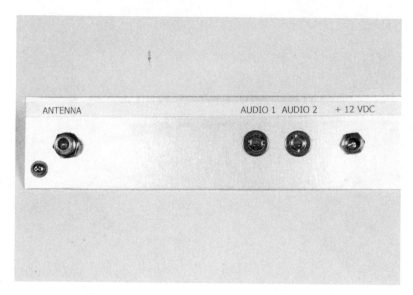

Figure 30–10 *Install the jacks and lug.*

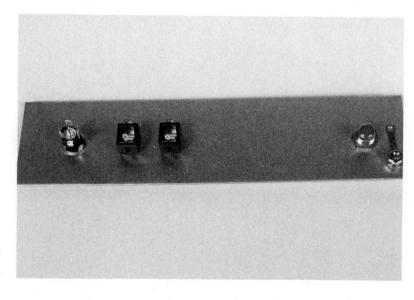

Figure 30–11 *Firmly secure the jacks and lug from the rear surface of the chassis rear panel.*

Figure 30–12 *The Radio JOVE receiver main circuit board.*

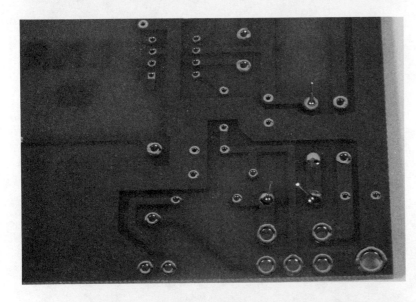

Figure 30–13 *Strive for making good, tight, professional-looking solder joints.*

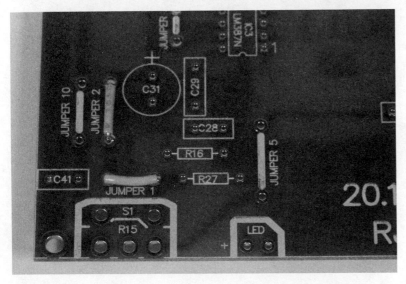

Figure 30–14 *Install wire jumpers.*

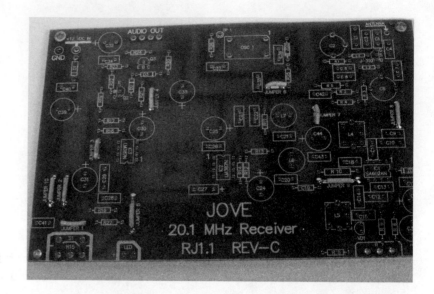

Figure 30–15 *Make sure all the jumpers are soldered into place without melting any of their plastic insulation.*

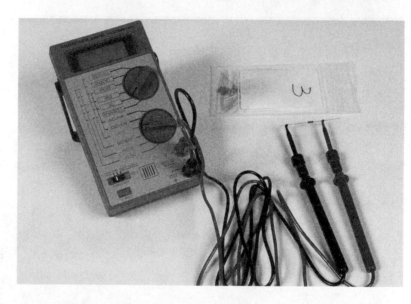

Figure 30–16 *Don't even try to match the metal film resistor color codes with the supplied resistors. Use a digital multimeter to determine the value for each resistor.*

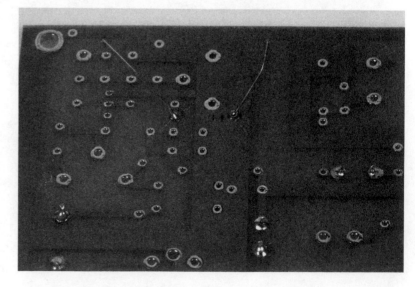

Figure 30–17 *Leave the resistor leads attached until you've thoroughly checked your soldering work.*

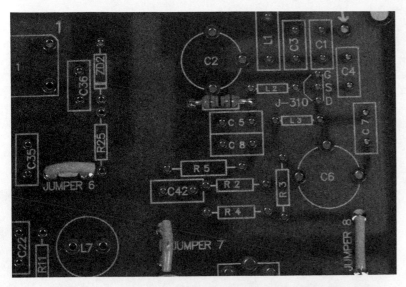

Figure 30–18 *Mount all resistors flush with the main circuit board.*

Figure 30–19 *All resistors have been installed.*

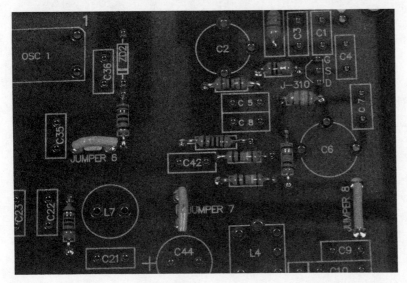

Figure 30–20 *Don't confuse an inductor (J-310) with a resistor.*

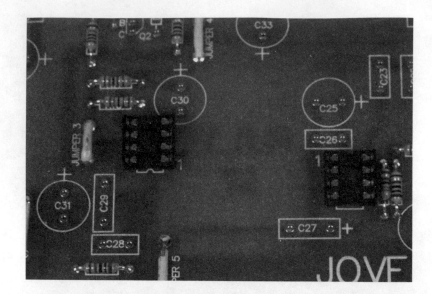

Figure 30–21 *Solder the IC chip sockets to the main circuit board.*

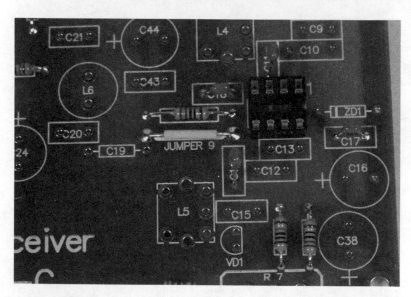

Figure 30–22 *Add the capacitors.*

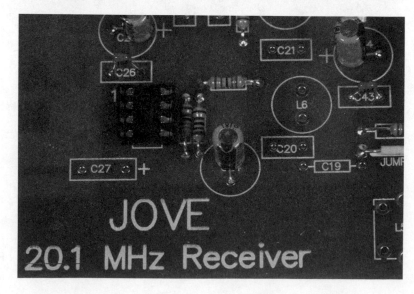

Figure 30–23 *Add the electrolytic capacitors. Watch the polarity. I recommend installing the positive (+) lead flush with the circuit board, while leaving the negative (–) lead running parallel with the circuit board.*

Figure 30–24   *Add the tuning capacitors.*

Figure 30–25   *Install the front switch and tuning potentiometer.*

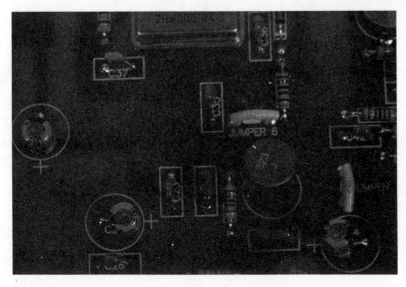

Figure 30–26   *This 20 MHz crystal is only used for tuning the receiver. Watch its orientation during installation.*

Figure 30–27 *These four output resistors are one of the two areas in this kit where you might want to install your own modification. Rather than soldering each resistor directly to the main circuit board, you might, instead, elect to attach each of them with a wire connector. This flexibility will be appreciated when you try to attach the chassis rear panel.*

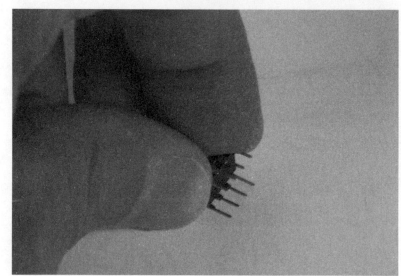

Figure 30–28 *Gently bend the pins of each IC to match its socket.*

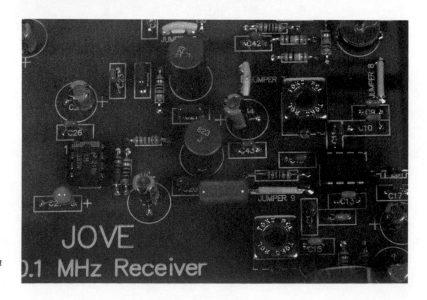

Figure 30–29 *Insert the IC chips according to pin 1 orientation and watch out for bent or misaligned pins.*

Figure 30–30 *Attach the chassis front panel to the main circuit board.*

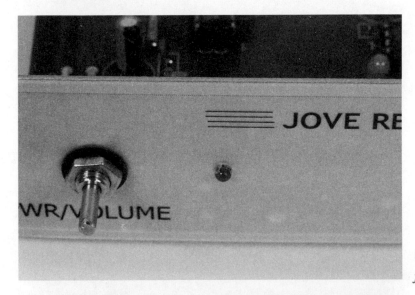

Figure 30–31 *Mount the LED to the front panel.*

Figure 30–32 *Solder the LED to the main circuit board.*

Figure 30–33 *Loosely attach the chassis left side panel with four channel guide spacers.*

Figure 30–34 *Be careful not to bend the two back channel guides.*

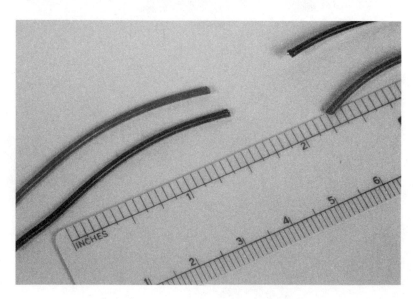

Figure 30–35 *Cut wire for the antenna and ground connections.*

Figure 30–36 *Solder the wire to the antenna and grounding lug.*

Figure 30–37 *Loosely attach the chassis right side panel.*

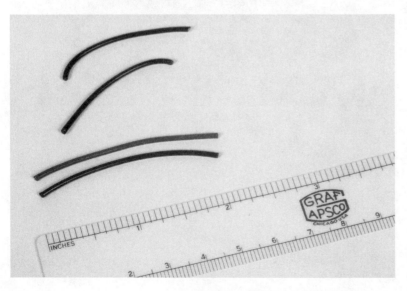

Figure 30–38 *Cut wire for the power connections.*

Figure 30–39 *Wire the power jack and the two audio output jacks. Now you can see the problem previously mentioned in Figure 30-27. The audio output resistors are wired directly to the rear panel and the main circuit board. Any movement of either the circuit board or the rear panel during the remaining assembly steps will shear these resistors off from their leads. Therefore, you wire connections and save yourself a headache later.*

Figure 30–40 *Here is the second area for user modification. The chassis top and bottom panels will not safely slide into the chassis channel guides. They were manufactured to the same width as the side panels and they will not fit. A foolproof installation modification is to slice off 3/8-inch from each panel. On the bottom panel just make sure you remove the metal from the side opposite the two spacer mounting holes.*

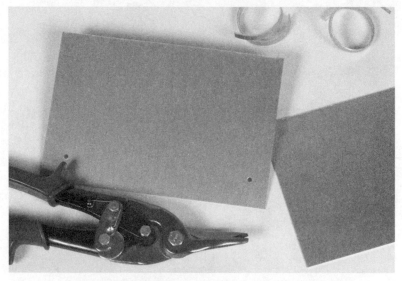

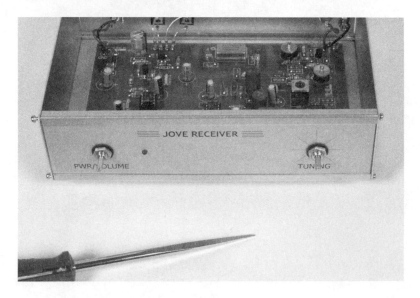

Figure 30–41 *Use a reamer or screwdriver for aligning holes in the chassis.*

Figure 30–42 *Temporarily clip a resistor across the antenna and ground lug jacks during the testing and tuning phase of the receiver assembly.*

Figure 30–43 *Test fit the chassis top panel.*

Figure 30–44 *When all tests and tuning steps are complete, disconnect the test crystal by cutting Jumper 6.*

Figure 30–45 *A Radio JOVE Receiver ready for listening to either Jupiter or the Sun.*

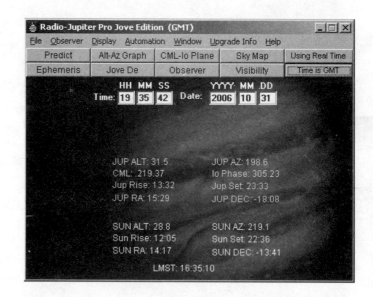

Figure 30–46 *The supplied Radio Jupiter Pro Jove Edition program is a PC-only program that can be used for setting up your Radio JOVE listening station.*

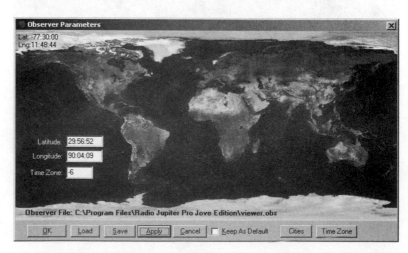

Figure 30–47 *You can change the software parameters for enabling anyone to participate in the Radio JOVE program from anywhere in the world.*

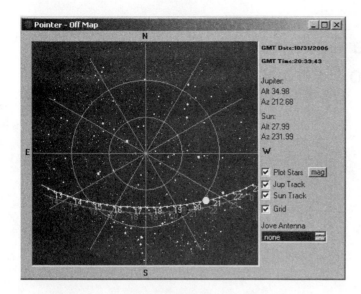

Figure 30–48 *The Radio Jupiter Pro Jove Edition program can also be used to locate Jupiter for visual observations.*

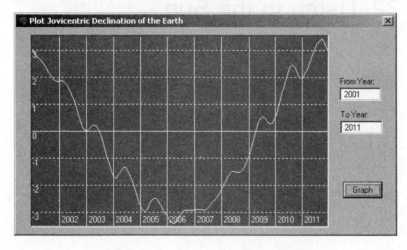

Figure 30–49 *Get an advanced heads-up for where Jupiter will be in, oh, say, 2010.*

Project 31. Using a Commercial Receiver for Radio Astronomy

What You Need:

- ICOM R-75 radio

Resources:

- NASA Radio JOVE Project— radiojove.gsfc.nasa.gov

If your hand can't comfortably hold a soldering iron (cough-cough), then opt for a commercial receiver. The Radio JOVE Project website lists a couple of different models that have been tested for making Jupiter observations.

Project 32. Listen to the Sun

What You Need:

- NASA JOVE radio receiver

Resources:

- NASA Radio JOVE Project— radiojove.gsfc.nasa.gov

When Radio JOVE Project members aren't listening to Jupiter, they reconfigure their antennas to listen to the sun. It's easier than you think, and this radio astronomy community has made some exciting observations of solar flares. Join the fun at the Radio JOVE Project website.

Project 33. How to Graph Your Star Sounds

What You Need:

- GarageBand

Resources:

- NASA Radio JOVE Project— radiojove.gsfc.nasa.gov

The recommended software for graphing your Radio JOVE Project observations is a PC-only program. Mac owners can join in the fun by using GarageBand to create a visual representation of your Jovian and solar observations. Just be aware that this only presents visual data, not scientific data.

Project 34. Group Listening

What You Need:

*   More Than One NASA Radio JOVE Site

Resources:

*   NASA Radio JOVE Project—
    radiojove.gsfc.nasa.gov

Radio astronomers from around the world join in with group observations with Radio JOVE Project. A special collaborative feature is built into the PC-only software that accompanies this project.

Project 35. Share Your Sounds with Podcasts

What You Need:

*   QuickTime Pro 7
*   iTunes

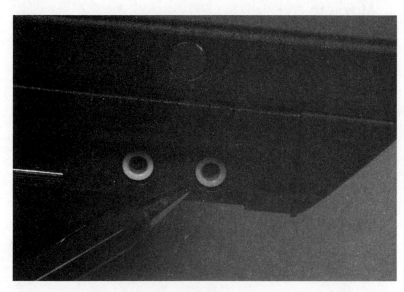

Figure 35–1   *Beat those PC JOVE users at their own game. Make a Mac podcast of your Radio JOVE recordings. Use the Line In port of your computer to jack into the Radio JOVE Receiver.*

New Player	⌘N
New Movie Recording	⌥⌘N
New Audio Recording	^⌥⌘N

Figure 35–2   *Use QuickTime 7 Pro to make a new audio recording.*

Figure 35–3 *Start the audio recording in QuickTime 7 Pro. During the recording, add commentary, music, and sound effects for a more complete podcast.*

Figure 35–4 *Your finished recording will be an audio.mov file.*

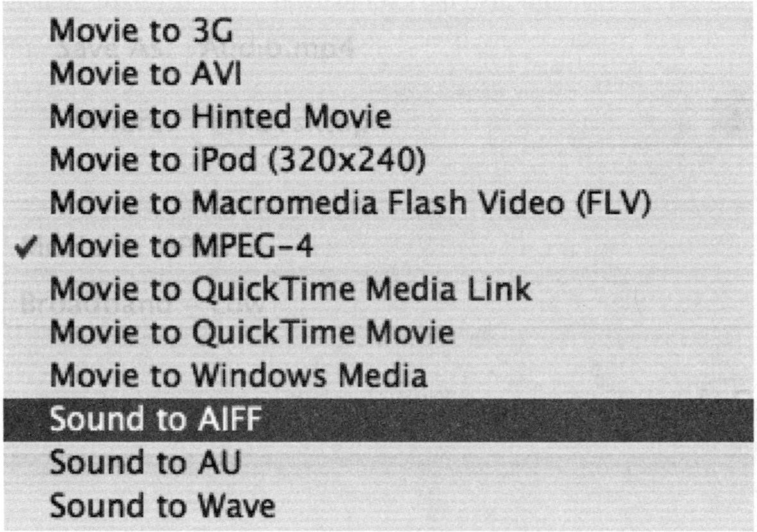

Movie to 3G
Movie to AVI
Movie to Hinted Movie
Movie to iPod (320x240)
Movie to Macromedia Flash Video (FLV)
✓ Movie to MPEG-4
Movie to QuickTime Media Link
Movie to QuickTime Movie
Movie to Windows Media
Sound to AIFF
Sound to AU
Sound to Wave

Figure 35–5 *Export the audio recording as an AIFF sound file.*

Open Stream...   ⌘U
Subscribe to Podcast...

Convert Selection to MP3

Figure 35–6 *Use iTunes to convert the AIFF sound file into an MP3. Then, publish your podcast with iTunes and share your Jovian mash-up with the rest of the universe.*

Chapter Five

On Your Mark, Get SETI

In the search for extraterrestrial intelligence, it's in everyone's best interest to participate in this discovery. And that's exactly the notion behind Search for Extraterrestrial Intelligence at home (SETI@home).

SETI@home is an organized search project that enables anyone, anywhere to help conduct the search for extraterrestrial intelligence. All you need is a personal computer and an online connection for participating in SETI@home. And, of course, you should also have a fanatical devotion to the discovery of something or someone who is truly "out of this world."

By installing and operating a free screensaver-like application, SETI@home participants combine their personal computing hardware with other SETI@home users and collectively scan the heavens for the "sounds of smart life."

You don't need your own radio telescope to be a SETI@home user, either. Registered members of this project receive data collected from the 1,000-foot diameter Arecibo Radio Telescope in Puerto Rico. Then, your computer does all the searching for you.

What will contact "look" like? Basically, it would be a "spike," which is the typical background "clutter" that fills most Arecibo data. This spike would be a narrow frequency band around 1420 MHz and it would last about 12 seconds.

Astonishingly, not everyone is totally sold on this concept of searching for extraterrestrial intelligence. In fact, the SETI program has led a storied life, especially while under the auspices of the U.S. Government.

Begun as a 1971 NASA program called "Project Cyclops," SETI was moved to the Jet Propulsion Laboratory (JPL) in 1979 for an "all-sky survey" called Microwave Observing Program (MOP). This project was soon cancelled by Senator William Proxmire in 1982. Later that year, it was resuscitated by Carl Sagan, and then officially "launched" in 1992 with a 34-meter radio telescope located at the Deep Space Communications Complex in the Mohave Desert. The project was renamed the High Resolution Microwave Survey (HRMS) in 1992, formally cancelled in 1993 by Senator Richard Bryan, transferred to the SETI Institute, and finally renamed Project Phoenix. Whew.

Now it's up to you.

Project 36. Going BOINC

What You Need:
- BOINC software
- Personal computer

Resources:
- BOINC: Compute for Science—boinc.berkeley.edu
- SETI@home—setiathome.berkeley.edu

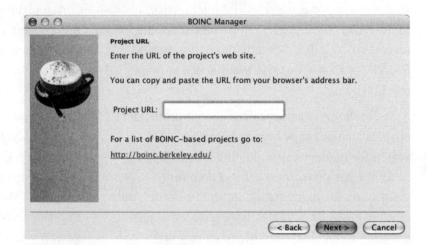

Figure 36–1   *Set up your Berkeley Open Infrastructure for Network Computing (BOINC) account.*

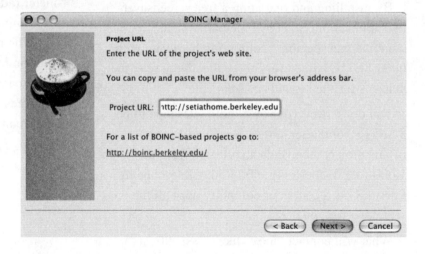

Figure 36–2   *Select a project you would like to support with your donated computer time.*

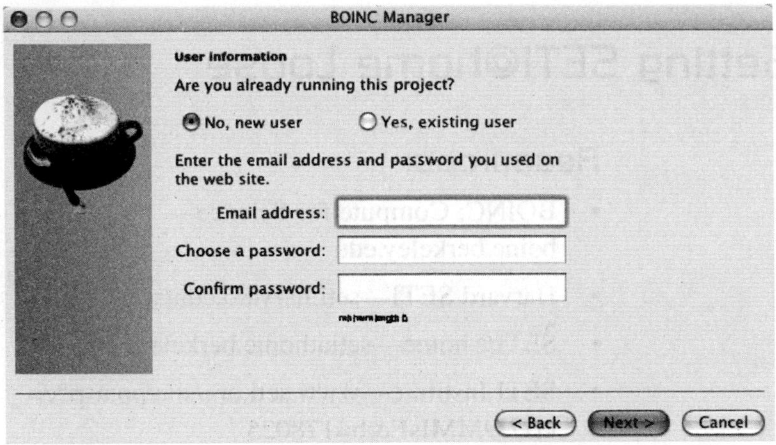

BOINC Manager

User information

Are you already running this project?

◉ No, new user   ○ Yes, existing user

Enter the email address and password you used on the web site.

Email address:

Choose a password:

Confirm password:

< Back   Next >   Cancel

Figure 36-3 *Put a human face (or at least a name) on your computer-donated time.*

Project	Application	Name	CPU time
SETI@home	setiathome_e..	13jn03aa.22301.3906.667316.3.85_0	---

Figure 36-4 *Get ready to compute.*

Project 37. Running SETI@home

What You Need:

- BOINC software
- Personal computer
- Network connection

Resources:

- BOINC: Compute for Science—boinc.berkeley.edu
- SETI@home—setiathome.berkeley.edu
- SETI Institute—www.seti.org/site/pp.asp?c=ktJ2J9MMIsE&b=178025
- SETI League—www.setileague.org/

Figure 37-1 *When you first use your BOINC Manager application, your selected project will download some application software to your computer. You will also receive your first packet of data to analyze.*

Project	File	Progress	Size	Elapsed Time	Speed
SETI@home	setiathome_5.13_powerpc-appl..	33.61%	0.92/2.72...	00:01:57	8.02 K
SETI@home	13jn03aa.22301.3906.667316....	9.04%	32.00/35...	00:00:45	0.73 K

Figure 37-2 *To dedicate maximum CPU time to your project, enable your screen-saver timer for its lowest value.*

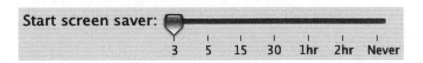

Start screen saver:

3 5 15 30 1hr 2hr Never

Project 38. Setting SETI@home Loose

What You Need:

- BOINC software
- Personal computer

Resources:

- BOINC: Compute for Science—boinc.berkeley.edu

- Harvard SETI—seti.harvard.edu/seti/

- SETI@home—setiathome.berkeley.edu

- SETI Institute—www.seti.org/site/pp.asp?c=ktJ2J9MMIsE&b=178025

- SETI League—www.setileague.org/

Figure 38–1 *Enable your computer's screen-saver function.*

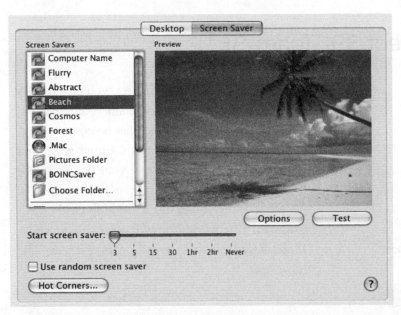

Figure 38–2 *Set your screen-saver's startup time for its lowest value. This will dedicate maximum CPU time to your project.*

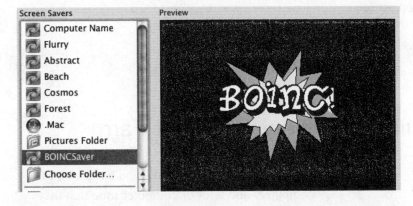

Figure 38–3 *Select the BOINC screen-saver application.*

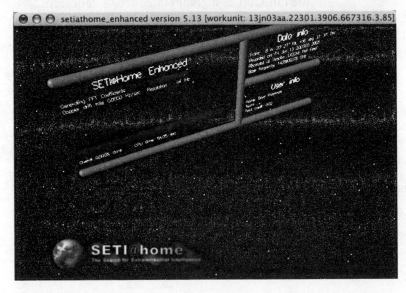

Figure 38–4 *This is what data from space really looks like.*

Project 39. Team SETI

What You Need:

- BOINC software
- Personal computer

Resources:

- BOINC: Compute for Science—
  boinc.berkeley.edu

- Search for Extraterrestrial Intelligence:
  The Planetary Society—
  www.planetary.org/explore/topics/seti/

- Harvard SETI—seti.harvard.edu/seti/

- SETI@home—setiathome.berkeley.edu

- SETI Institute—www.seti.org/site/pp.asp?c=
  ktJ2J9MMIsE&b=178025

- SETI League—www.setileague.org/

When you join the SETI project, you can specify that you are a member of a "team." Actually, either you or your computer can be considered the true "member" of this team. Either way, your collaborative efforts will have your donated hours logged into a lump sum for the whole team. The people at BOINC call this a Desktop Grid Computing network. You will need a dedicated Linux-based network server for this project, however.

Project 40. Building a Giant SETI Ant Farm

What You Need:

- BOINC software
- Personal computer
- Several other SETI friends

Resources:

- BOINC: Compute for Science—
  boinc.berkeley.edu

- Search for Extraterrestrial Intelligence:
  The Planetary Society—
  www.planetary.org/explore/topics/seti/

- Harvard SETI—seti.harvard.edu/seti/

- SETI@home—setiathome.berkeley.edu

- SETI Institute—www.seti.org/site/pp.asp?c=
  ktJ2J9MMIsE&b=178025

- SETI League—www.setileague.org/

Do you have a lot of friends who have a lot of computers and a lot of computer time that they would like to donate to the SETI project? Well, then link up as a giant collective or "ant farm." The BOINC folks call this a Virtual Campus Supercomputing Center (VCSC). But I like ant farm better. You need a server that runs Linux, however, for enabling a VCSC.

Chapter Six

You Can't Tell the Stars Without a Map

One evil genius's "beetle juice" is another evil genius's Betelgeuse (pronounced *betl-jooz*). *Betelgeuse* is a red supergiant star with a variable intensity exhibiting a period of change lasting approximately seven years. It sits in the shoulder of the famed constellation Orion, directly opposite Bellatrix.

While almost every evil genius knows this little bit of stellar star stuff, not many can finger where those funky names came from and who coined them.

The bulk of our named stars are derived from the Arabs and Greeks. In the case of Betelgeuse, the derivation is attributed to an Arabic translation of "house of the twins." This odd-sounding translation is due to the Arab astronomers associating Betelgeuse with the nearby Gemini constellation. Pretty nifty, eh?

In the eighteenth century, Royal Astronomer Reverend John Flemsteed assigned numbers to his listing of stars based on the 48 primary constellations. According to Flemsteed's catalog, Betelgeuse is known as "58 Orionis."

An earlier attempt at naming the stars in constellations was undertaken in 1603 by Johann Bayer. His naming convention was published in a star catalog called *Uranometria*. According to Bayer's star catalog, Betelgeuse was named Alpha Orionis, the "alpha" designation serving as a label for star brightness, with alpha being the brightest star in any particular constellation. Bayer's name is dubious, however. Remember, Betelgeuse is a variable star. Therefore, it's brightness changes over a seven-year period and Bayer must've recorded his observations near the peak of the Betelgeuse cycle.

Just like these astronomers of yesteryear, other scientists have published star catalogs that have added to our naming conventions. There's the Messier Catalogue (published in 1781; listing 110 objects), New General Catalog and two Index Catalogues (NGC and IC, respectively; published in 1888; listing over 13,000 objects), Smithsonian Astrophysical Observatory catalog (SAO; published in 1966; listing over 258,000 objects), and the Hubble Space Telescope catalog (HST; published in 1993; listing over 19 million objects).

How many objects can you list in your catalog?

What You Need:

- Just you
- Carpenter's spirit level
- Night Owl Optics® Night Vision Scope

Resources:

- Night Owl Optics—www.nightowloptics.com

Figure 41–1 *A small reliable spirit level is vital for establishing a useable telescope mount.*

Figure 41–2 *Ensure that the telescope's tripod mount is level before you begin to make any adjustments for declination or right ascension.*

Figure 41-3 *A Night Owl Optics Night Vision scope can be used for locating visual obstructions that might be blocking your telescope's field of view.*

Project 42. How to Figure Degrees with Your Fingers

What You Need:

* Just you

Figure 42-1 *Hold your thumb at arm's length and you can cover about a 2° field of view.*

Figure 42–2 *A closed fist covers about a 10° field of view.*

Figure 42–3 *An open hand held at arm's length can cover a whopping 20° field of view.*

Project 43. Create Your Own Star Map

What You Need:

- Your eyes
- Zhumell Aurora 70 telescope

Resources:

- The Hubble Space Telescope Data Archive—www.adass.org/adass/proceedings/adass94/bornek.html
- Multimission Archive at Space Telescope (MAST) HST—archive.stsci.edu/hst/
- National Space Science Data Center—nssdc.gsfc.nasa.gov/
- Smithsonian Astrophysical Observatory Star Catalog—heasarc.gsfc.nasa.gov/W3Browse/star-catalog/sao.html
- Zhumell, Inc.—www.zhumell.com

Making your own Star Map is easier than you might think. For example, find a star or other celestial object. Center it in your telescope. Now, record all the data for altitude, azimuth, right ascension, and declination. Also, don't forget to record the date, time, latitude, and longitude for each object observation.

Project 44. Manage the Pleiades

What You Need:

- Starry Night Pro
- Zhumell Tachyon 25×100 astronomical binoculars

Resources:

- Imaginova®—www.imaginova.com
- Starry Night Store—www.starrynight.com
- Zhumell, Inc.—www.zhumell.com

Typically rising an hour before the entire Taurus constellation is visible, the Pleiades or Seven Sisters (or M45) is a gorgeous blue star cluster about 400 light years from Earth. On the first of the year, in the northern hemisphere, you can generally find the Pleiades at 24°, 3h 48m.

Project 45. How to Find a Tempest in a Teapot

What You Need:

- Zhumell Tachyon 25×100 astronomical binoculars

Resources:

- Zhumell, Inc.—www.zhumell.com

Inside the constellation Sagittarius is a curious star formation. Some astronomers claim it looks like a teapot. What do you think it looks like? First, find Sagittarius. On a late summer night in the northern hemisphere, you can find the heart of Sagittarius at –30°, 18h 30m. Inside this heart is where you can see a teapot. In fact, M54, M70, and M69 form the base of this teapot.

Project 46. Create Your Own Full-Sky Star Charts

What You Need:

- Starry Night Pro
- Zhumell Tachyon 25×100 astronomical binoculars

Resources:

- The Hubble Space Telescope Data Archive— www.adass.org/adass/proceedings/adass94/ bornek.html
- Smithsonian Astrophysical Observatory Star Catalog—heasarc.gsfc.nasa.gov/W3Browse/ star-catalog/sao.html
- Starry Night Store—www.starrynight.com
- Zhumell, Inc.—www.zhumell.com

A good pair of astronomy binoculars fixed on a steady tripod are the best aid for creating your own star chart. Begin your star-charting exercises by making your own observations of a well-known star group. For example, choose the Big Dipper or Orion. Now, armed with paper, pencil, and a red flashlight (for example, the Great Red Spot Junior Red LED Shake Flashlight), start drawing your own star chart. Include as many references (for example, position, attitude, date, time, and magnitude) for each star as you can see with your binoculars. Then, use your naked eyes to fill in any blanks in your star chart. Show the completed chart to someone else and have them try to name the star group. Consider yourself a 100 percent evil genius if that certain someone correctly names this star group.

Project 47. Learn How to "Map" Your Telescope

What You Need:

- Zhumell Aurora 70 telescope

Resources:

- Zhumell, Inc.—www.zhumell.com

Figure 47–1 *A red flashlight like the Great Red Spot Junior Red LED Shake Flashlight is ideal for reading telescope markings without compromising your night vision.*

Project 48. Decipher Constellations

What You Need:

- Just you

Resources:

- The Hubble Space Telescope Data Archive—www.adass.org/adass/proceedings/adass94/bornek.html

- Smithsonian Astrophysical Observatory Star Catalog—heasarc.gsfc.nasa.gov/W3Browse/star-catalog/sao.html

Pick a star constellation that doesn't have a stick figure drawing around its stars. Now, see if you can envision what the early astronomers thought this constellation looked like. What were they thinking? They had some very vivid imaginations, didn't they?

Project 49. White Dwarfs, Supernovae, and Black Holes

What You Need:

- Starry Night Pro
- Zhumell Tachyon 25×100 astronomical binoculars

Resources:

- Starry Night Store— http://www.starrynight.com
- Zhumell, Inc.—www.zhumell.com

Just what is a black hole? Can you see a black hole? Well, way up near the head of Scorpius (inside its left pincer) resides a dense binary star. Known as X-1 (isn't that a great name for a black hole?), one of the stars is being influenced by another "invisible" star—its binary. Some astronomers speculate this invisible companion could be a black hole. In midsummer in the northern hemisphere, you can find X-1 at –16°, 16h 20m.

Project 50. Pick a Ride on the Tail of a Comet

What You Need:

- Uncle Milton Star Theater 2 with Meteor Maker™

Resources:

- Uncle Milton Toys—www.unclemilton.com

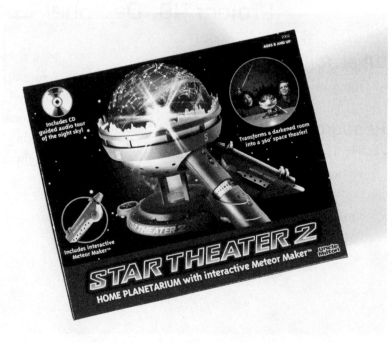

Figure 50–1 *Uncle Milton Star Theater® 2.*

Figure 50–2   *The Star Theater 2 base is marked with compass directions and a simulated city skyline.*

Figure 50–3   *Orient the built-in compass with north.*

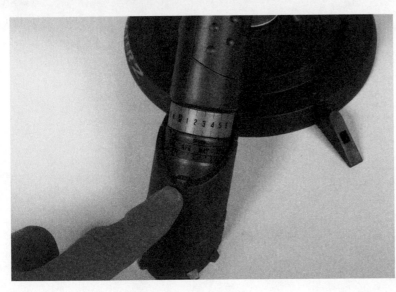

Figure 50–4   *Turn the projection wand's dial until the desired month is lined up with the indicator mark.*

Figure 50–5 *The internal lamp can either project the constellations on a ceiling or wall, or it can be used for charging the phosphorescent stars, enabling them to glow in the dark.*

Figure 50–6 *A dark room is ideal for viewing the Star Theater 2 planetarium show.*

Chapter Seven

One Eye on the Sky

Want to start a fistfight between a couple of astronomers? Ask them who invented the telescope. In less time than it takes to pose this question, you might have two distinctly different opinions: Hans Lipperhey vs. Galileo Galilei.

Actually, both answers *could* be correct. Why? One astronomer was answering your question based on technical merits, while the other was answering it on a philosophical basis. Huh? OK, here goes.

The telescope was "invented" in 1608 by one or several different simultaneous tinkerers. Two Dutch experimenters, Hans Lipperhey and Jacob Metius, demonstrated examples of their "inventions" at The Hague. Another Dutch experimenter, Sacharias Janssen, made a similar optical discovery around that same time.

Elsewhere in Europe, an inventor in Italy made his own telescope in 1609. The purpose for this telescope was for observing the sky around the Earth. This Italian was Galileo. And, with his telescope, Galileo made numerous astronomical observations, including the moons of Jupiter and nearby nebulae.

Now, if one of your combatants answered your question with Sir Isaac Newton, then you have another correct answer. Sort of.

Not content with the size limitations of Galileo-type refracting telescopes, Newton built a telescope that used a mirror for gathering and reflecting light to another mirror. This second mirror then reflected the light into an eyepiece. This invention, which was completed in 1671, was the reflecting telescope.

So there has been well over 300 years' worth of telescope viewing at the heavens. What are you waiting for?

Resources:

- The Galileo Project—galileo.rice.edu

Project 51. Ahoy Mates, I Spy a Star

What You Need:

- Konus KJ-8

Resources:

- Konus—www.konus.com

Figure 51–1   *Konus KJ-8 telescope.*

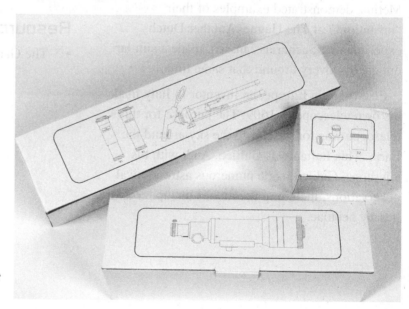

Figure 51–2   *Make no mistake, this is not a professional-grade telescope. The KJ-8 is an entry-grade telescope that balances minimal settings and lightweight construction against very good optics.*

Figure 51–3   *Mount the scope on its tripod.*

Figure 51–4   *Attach the vertical movement adjustment handle.*

Figure 51–5   *An eyepiece can be mounted directly on the focusing tube.*

Project 51. Ahoy Mates, I Spy a Star

Figure 51–6 *A diagonal enhances the viewing experience and sharpens the focusing adjustment.*

Figure 51–7 *An image-erecting eyepiece.*

Figure 51–8 *A 3x Barlow lens.*

Figure 51–9 *A protective lens cap.*

Figure 51–10 *Ready for viewing or using as a spotting scope.*

Figure 51–11 *This setup provides reasonably good images, but at the price of very sensitive hardware and a delicate tripod.*

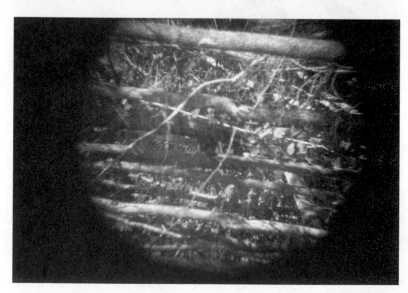

Figure 51–12 *You can easily snap an image with a digital camera through the eyepiece of the KJ-8.*

Project 52. How to Assemble a Refractor Telescope

What You Need:
- Zhumell Aurora 70

Resources:
- Zhumell, Inc.—www.zhumell.com

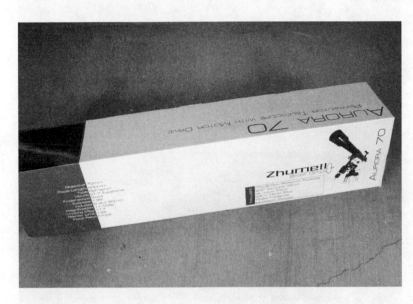

Figure 52–1   *Zhumell Aurora 70.*

Figure 52–2   *The Aurora 70 features a 70mm objective with a 900mm focal length, uses 1.25-inch eyepieces, and comes with a right ascension motor-drive mechanism.*

Figure 52–3 *Adjust all three tripods legs to the same length.*

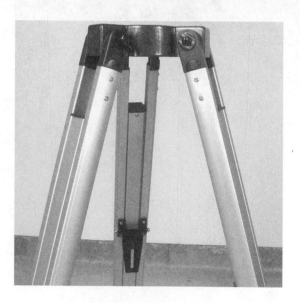

Figure 52–4 *Attach the legs to the tripod mount.*

Figure 52–5 *Carefully lift the tripod up with all three legs firmly positioned on the ground.*

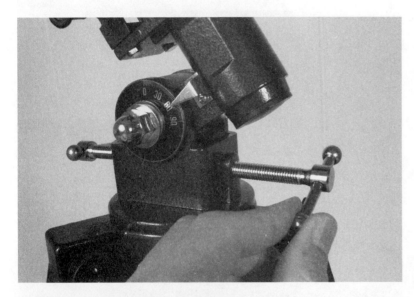

Figure 52–6 *Attach the telescope mount to the tripod mount.*

Figure 52–7 *Loosen the altitude block.*

Figure 52–8 *The motor drive is attached to these two points.*

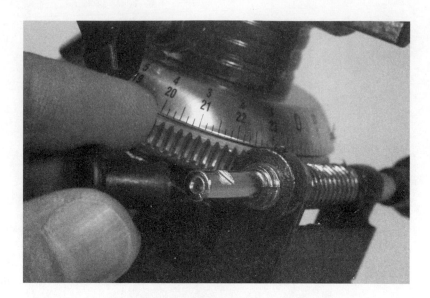

Figure 52–9 *This right ascension shaft is used for connecting the motor drive to the telescope.*

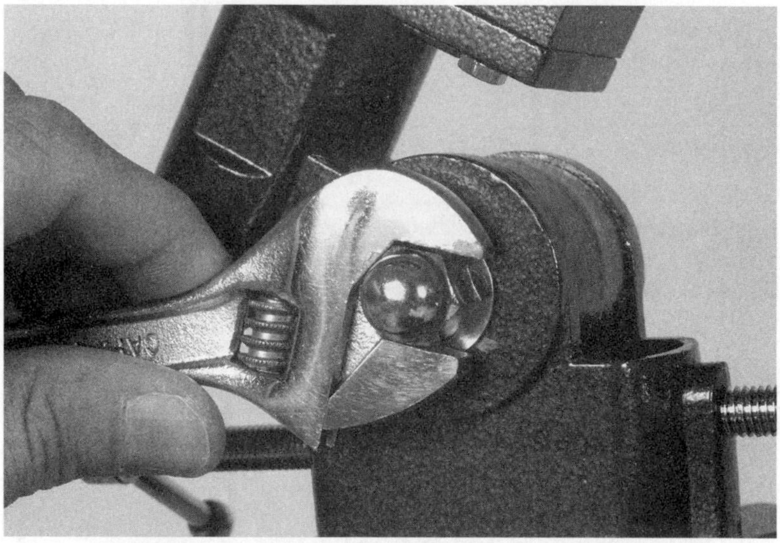

Figure 52–10 *Loosen this nut for holding the motor drive.*

Figure 52–11 *Attach the motor-drive mounting plate.*

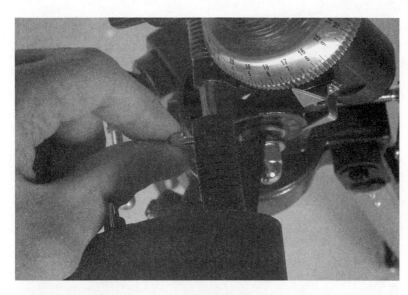

Figure 52–12 *Connect the motor drive shaft to the right ascension shaft.*

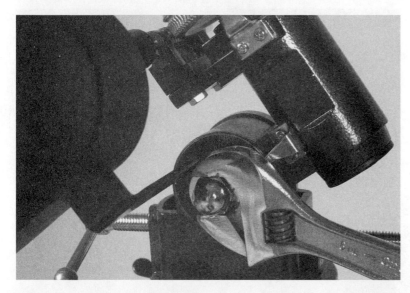

Figure 52–13 *Reattach the nut from Figure 52-10 and securely fix the motor drive to the telescope mount.*

Figure 52–14 *Prepare the counterweight.*

Figure 52–15 *Attach the counterweight to the telescope mount.*

Figure 52–16 *Remove the wing nuts from the telescope saddle.*

Figure 52–17 *Attach the saddle to the telescope mount.*

Figure 52–18   *Gently lay the telescope tube in the saddle.*

Figure 52–19   *Slide the tube to a balanced point and tighten the saddle.*

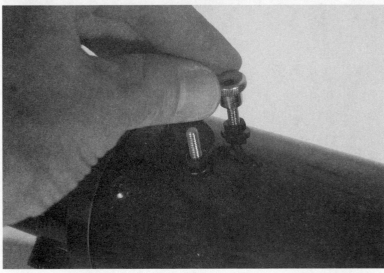

Figure 52–20   *Remove the mounting nuts for the finder.*

Figure 52–21 *Attach the finder to the telescope tube, open the saddle again, and readjust the telescope's balance.*

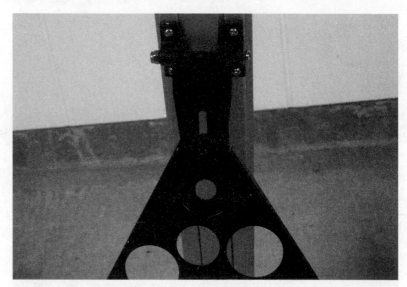

Figure 52–22 *Attach the accessory tray to one of the tripod's legs.*

Figure 52–23 *Loosely attach the remaining tripod legs to the accessory tray.*

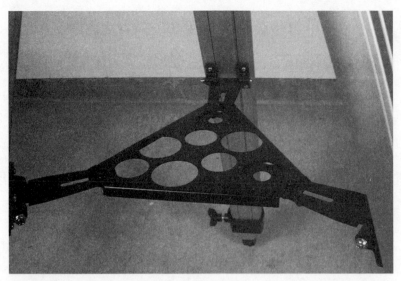

Figure 52–24 *Check the level and balance of the tripod, and then tighten the accessory tray screws.*

Figure 52–25 *The telescope is now ready to receive its optics.*

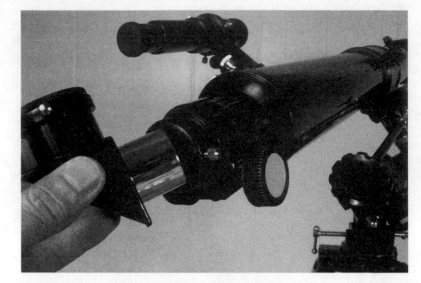

Figure 52–26 *Attach the diagonal to the telescope's focusing tube.*

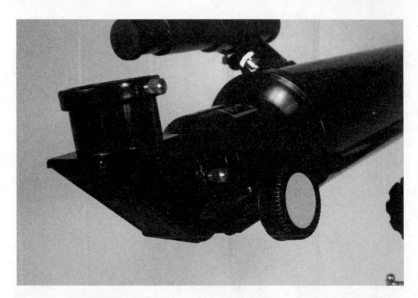

Figure 52–27 *This diagonal can use 1.25-inch eyepieces.*

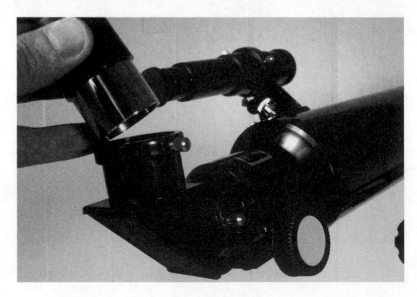

Figure 52–28 *Slip an eyepiece into the diagonal.*

Figure 52–29 *Tighten the eyepiece-retaining set screw.*

Figure 52–30 *Now get out there and see something.*

Project 53. How to Assemble a Reflector Telescope

What You Need:

- Konus Konusmotor-500 reflector telescope

Resources:

- Konus—www.konus.com

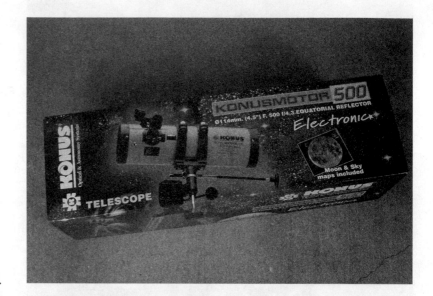

Figure 53–1   *Konus Konusmotor-500.*

Figure 53–2   *This is a highly portable, extremely powerful 4.5-inch Newtonian reflector telescope that accepts 1.25-inch eyepieces, uses a right ascension motor drive, and features a unique red laser dot "stardot" finder.*

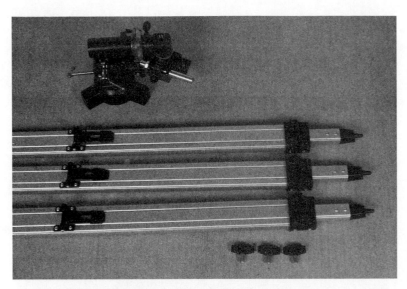

Figure 53–3 *The parts for the tripod mount.*

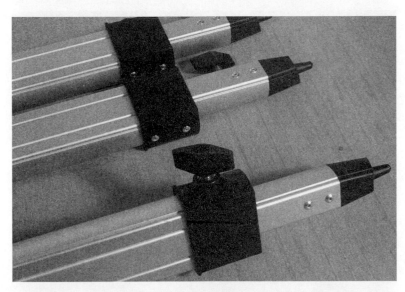

Figure 53–4 *Each tripod leg includes a sliding leg extender.*

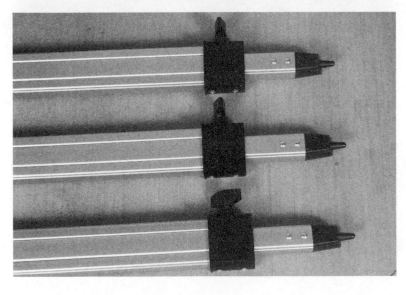

Figure 53–5 *Extend all three tripod legs to the same length.*

Figure 53–6 *Use bolts and wing nuts for attaching each tripod leg to the tripod mount.*

Figure 53–7 *Make sure you orient the wing nuts for easy access to the telescope mount's rotation set screw.*

Figure 53–8 *Check all the other telescope mount screws for clearance around the tripod leg mounting wing nuts.*

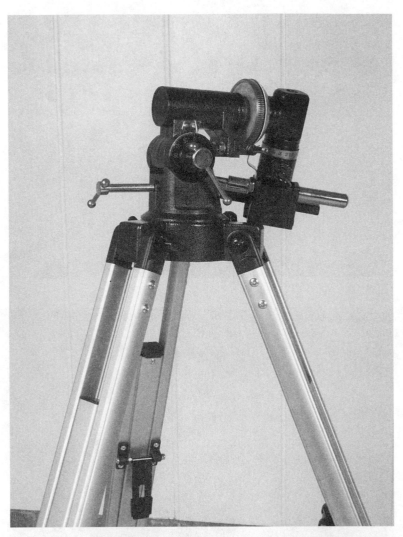

Figure 53–9 *Raise the tripod and level the legs.*

Figure 53–10 *Loosen the altitude clamp.*

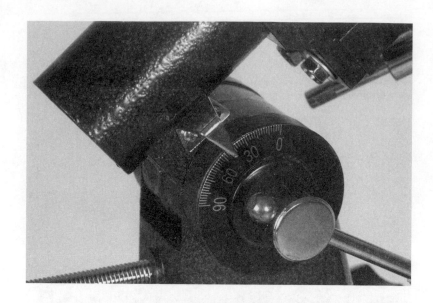

Figure 53–11   *Elevate the telescope mount to approximately 45°.*

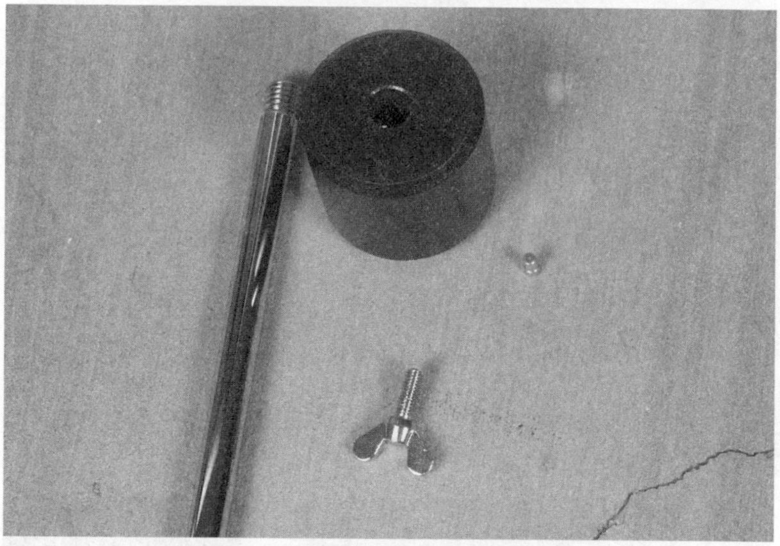

Figure 53–12   *Prepare the counterweight parts.*

Figure 53–13   *Assemble the counterweight.*

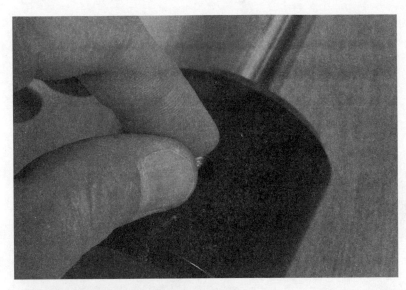

Figure 53–14 *Make sure you insert the counterweight set screw in the proper orientation. If it is backward, the weight will slide loosely up and down the counterweight mounting shaft.*

Figure 53–15 *Tighten the blocking screw for the declination adjustment.*

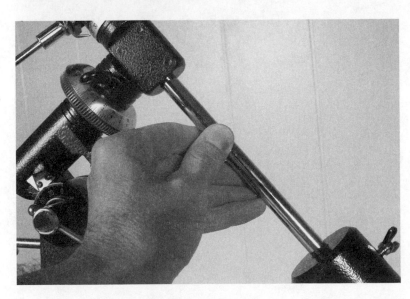

Figure 53–16 *Install the counterweight.*

Figure 53–17 *Attach the short flexible cable to the declination worm drive.*

Figure 53–18 *Attach the long flexible cable to the right ascension worm drive.*

Figure 53–19 *Loosely attach one tripod leg to the accessory tray.*

Figure 53–20 *Don't tighten this mounting bolt, yet.*

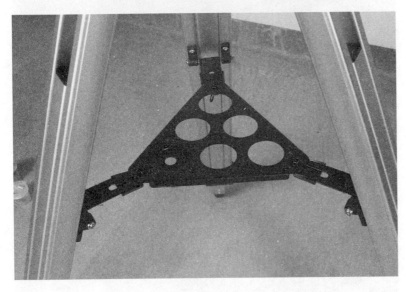

Figure 53–21 *Loosely attach the remaining two tripod legs to the accessory tray.*

Figure 53–22 *Level both the tripod and telescope mounts, and then tighten the accessory tray bolts.*

Figure 53–23 *Attach the telescope saddle to the telescope mount.*

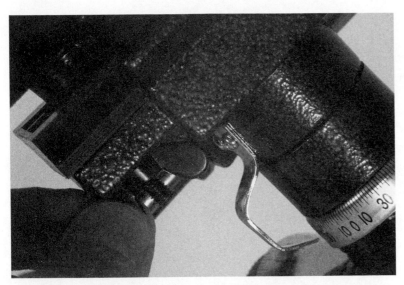

Figure 53-24 *Watch the orientation of the saddle's wing nuts to ensure clearance is in all movements of operation.*

Figure 53-25 *Open the saddle.*

Figure 53-26 *Gently lay the telescope in the saddle.*

Figure 53–27 *Check the balance of the telescope, and then tighten the saddle's clamp.*

Figure 53–28 *No diagonals are needed with a reflector telescope. Eyepieces are inserted directly into the focusing tube.*

Figure 53–29 *A set screw will hold the eyepiece in place.*

Figure 53–30 *Attach the finder to the telescope tube.*

Figure 53–31 *Carefully remove the lens cap. If you can't grasp the lens cap, remove the center portion for easier gripping.*

Figure 53–32 *Never touch this spider. This assembly holds a front-surface mirror that is used for channeling the main mirror's image into the eyepiece.*

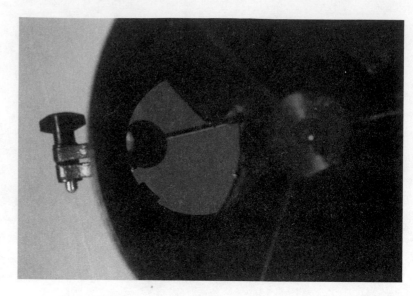

Figure 53–33 *This is a professionally polished mirror that should be protected from moisture, dust, and abrasions.*

Project 54. How to Assemble a Dobsonian Telescope

What You Need:

- Zhumell 10-inch reflector telescope with an altitude/azimuth mount: commonly called a Dobsonian type

Resources:

- Zhumell, Inc.—www.zhumell.com

Figure 54–1 *Begin the assembly of the Zhumell 10-inch Dobsonian telescope with the base.*

Figure 54–2 *Dobsonian mount telescopes are currently the rage among amateur astronomers. Among the best in this line, the Zhumell 10-inch Dobsonian accepts either 1.25- or 2-inch eyepieces in its 2-inch Crayford Focuser. It also has a built-in cooling fan for the main mirror, a laser collimator, and an 8×50 finder.*

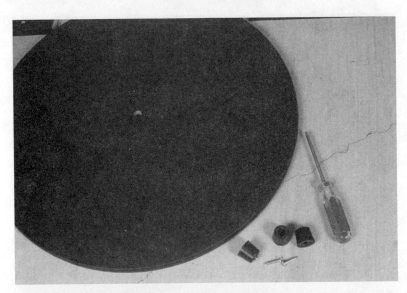

Figure 54–3 *Attach three rubber feet to the telescope mount base.*

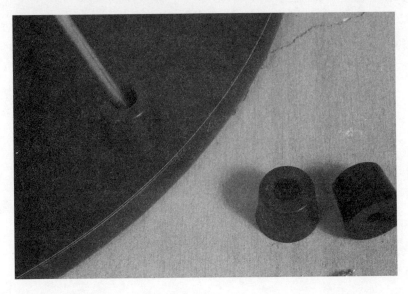

Figure 54–4 *Do not overtighten the screws holding the rubber feet.*

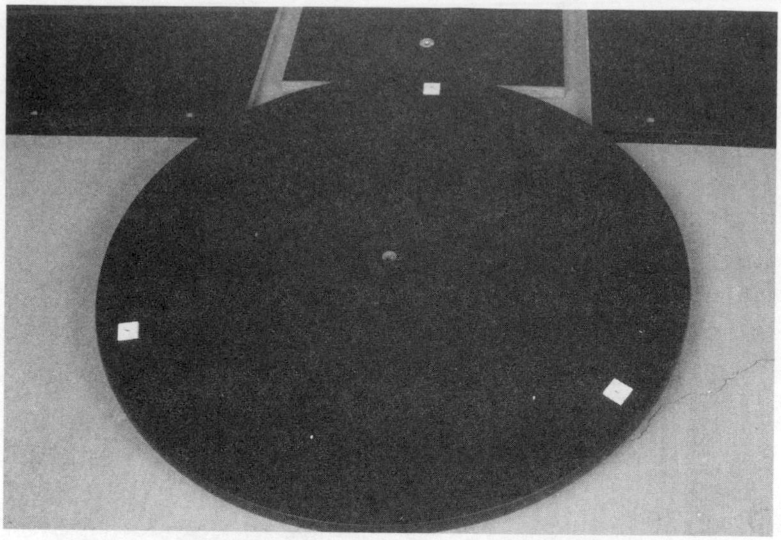

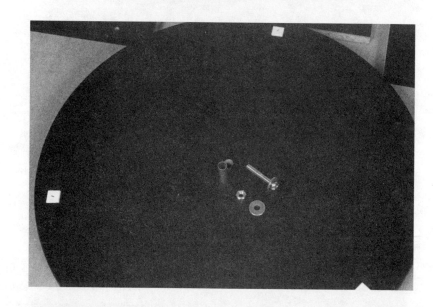

Figure 54–5 *Set the base on its rubber feet.*

Figure 54–6 *The parts of the base-center rotating mount.*

Figure 54–7 *Loosely install the center rotating mount.*

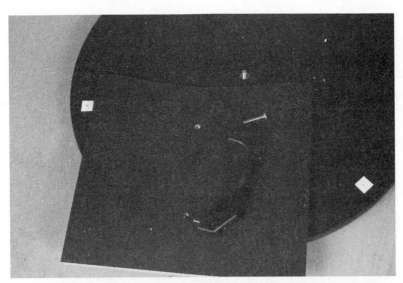

Figure 54–8 *Mount the handle to the base front.*

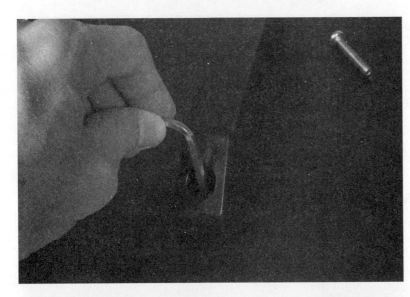

Figure 54–9 *Do not overtighten these bolts.*

Figure 54–10 *Lay the rotating plate on the mount base.*

Figure 54–11   *Insert the mounting screws in the rotating plate for holding the base sides.*

Figure 54–12   *Loosely attach these screws until all sides have been assembled to the rotating plate.*

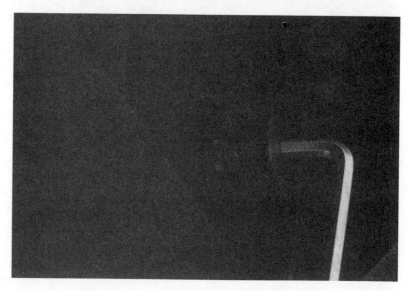

Figure 54–13   *Do not overtighten these screws.*

Figure 54–14   *Leave each screw head slightly exposed.*

Figure 54–15   *After all three sides are loosely assembled, then begin tightening each screw.*

Figure 54–16   *Mount the rotating plate on the mount base with the center rotating mount.*

Figure 54–17 *Attach the accessory tray to the base side panel.*

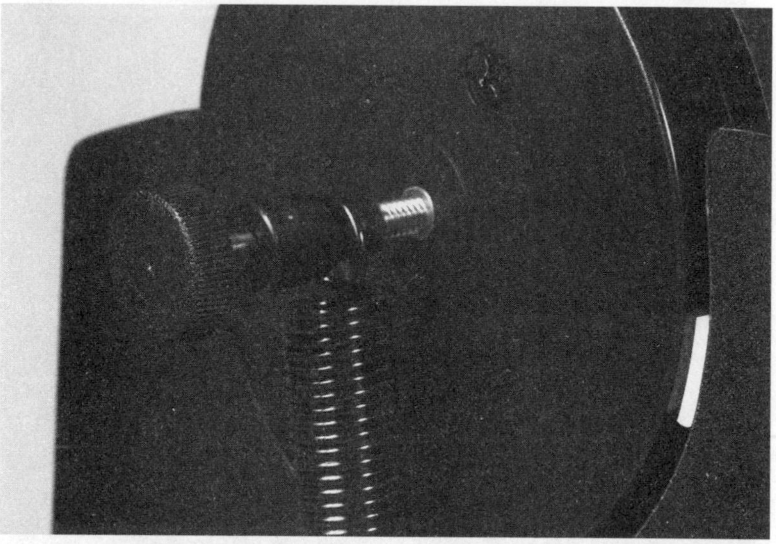

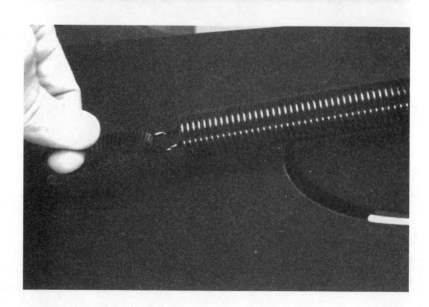

Figure 54–18 *Gently set the telescope tube in the mount. Thread the spring mount into the pivot posts.*

Figure 54–19 *Stretch the spring by pulling on the lanyard.*

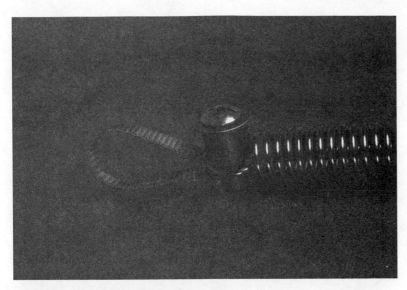

Figure 54–20 *Fix the stretched spring to each base side panel.*

Figure 54–21 *Raise and rotate the telescope tube for easy access to the focuser.*

Figure 54–22 *The Crayford Focuser will accept either 1.25- or 2-inch eyepieces.*

Figure 54–23 *Attach the finder mount and insert the finder.*

Figure 54–24 *Remove the 1.25-inch eyepiece mount for installing a 2-inch eyepiece in the focuser.*

Figure 54–25 *This telescope has incredible viewing power. A nice touch is the fine focus knob on the focuser.*

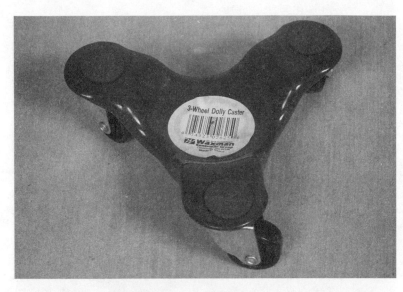

Figure 54-26 *One drawback of the Dobsonian telescope is that it can be unwieldy to move. A mover's dolly is a great tool for moving your "Dob" about and can be found in most DIY hardware centers for less than $10.*

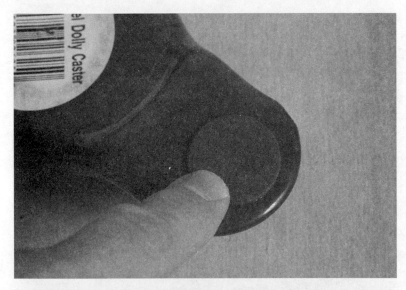

Figure 54-27 *Just make sure your dolly has friction pads for keeping your dolly underneath your Dob.*

Figure 54-28 *Inside the Dobsonian mount is a great cubbyhole for storing your dolly.*

Figure 54–29 *A mirror cooling fan is built into the base of the Zhumell Dobsonian telescope tube.*

Project 55. Track Stars with Clock Drive

What You Need:

- Zhumell Aurora 70

Resources:

- Zhumell, Inc.—www.zhumell.com

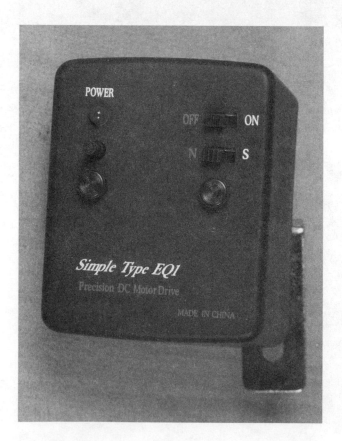

Figure 55–1 *Konus Konusmotor-500 right ascension (RA) motor drive.*

Figure 55–2 *The Konus motor drive has two attachment points: the RA worm drive shaft and the vertical adjustment mounting bolt.*

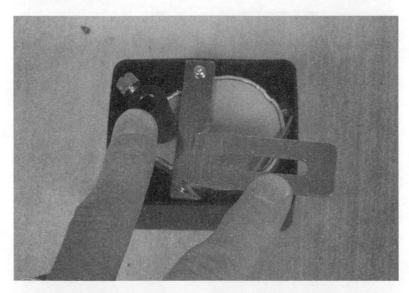

Figure 55–3 *The corresponding attachment points on the motor drive.*

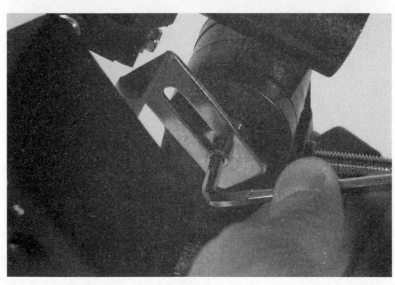

Figure 55–4 *Attach the motor drive to the vertical adjustment bolt.*

Figure 55–5 *Connect the motor drive shaft to the RA worm drive.*

Figure 55–6 *Remove the motor drive cover.*

Figure 55–7 *Connect one 9V battery to the motor drive.*

Figure 55–8 *Install the battery and replace the motor drive cover.*

Figure 55–9 *Zhumell Aurora 70 RA motor drive.*

Figure 55–10 *The Zhumell motor drive uses two 9V batteries.*

Figure 55–11 *Reattach the motor drive cover and connect the motor's shaft to the RA worm drive.*

Project 56. When Two Eyes Are Better Than One

What You Need:

- Zhumell Tachyon 25×100 astronomical binoculars

Resources:

- Zhumell, Inc.—www.zhumell.com

Figure 56–1 *While these binoculars are ideal for sports and nature viewing, they might not match your astronomical viewing requirements. What you need are astronomical binoculars.*

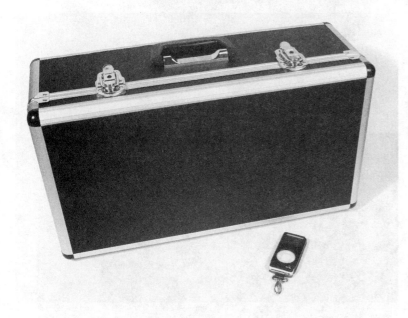

Figure 56–2 *Zhumell Tachyon 25×100 Astronomical Binoculars.*

Figure 56–3 *Now that's a binocular. These all-metal, rugged, waterproof binoculars are able to accept 1.25-inch astronomical filters.*

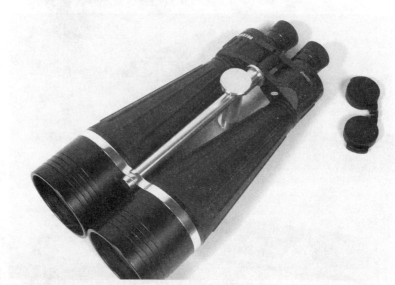

Figure 56–4 *A tripod adapter is included with these Zhumell binoculars.*

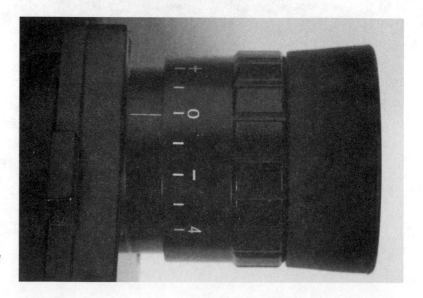

Figure 56–5 *Each ocular is gradated for precision adjustment.*

Figure 56-6 *If you have a strong-enough neck, you can attach the included neck strap.*

Figure 56-7 *Unlike conventional binoculars, these astronomical binoculars are focused with the oculars. Each eyepiece must be individually focused.*

Figure 56-8 *Both lenses are multicoated.*

Project 57. Finding Celestial Bodies with a Finder

What You Need:

- Zhumell 10-Inch Dobsonian reflector telescope

Resources:

- Zhumell, Inc.—www.zhumell.com

Figure 57–1 *A Zhumell 8×50 finder scope is an excellent choice for a Dobsonian telescope.*

Figure 57–2 *A rubber O ring is used for ensuring a snug fit inside the finder mount.*

Figure 57–3 *Slip the O ring into the groove on the finder scope's body tube.*

Figure 57–4 *Ready for mounting on a Dobsonian telescope.*

Figure 57–5 *Use a collimator for aligning the mirrors in Dobsonian and Newtonian telescopes.*

Figure 57–6   *Add filters to your eyepieces for enhancing your observations.*

Figure 57–7   *Size does make a difference: a 2-inch eyepiece (left) and a 1.25-inch eyepiece (right).*

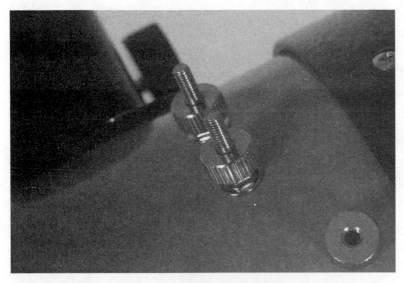

Figure 57-8 *Most telescopes are predrilled for accepting a finder scope.*

Figure 57-9 *Secure your finder scope to your telescope tube.*

Figure 57-10 *Focus on a terrestrial object (approximately 1,000 yards distant) with the main telescope. Then, center the same object in the finder scope.*

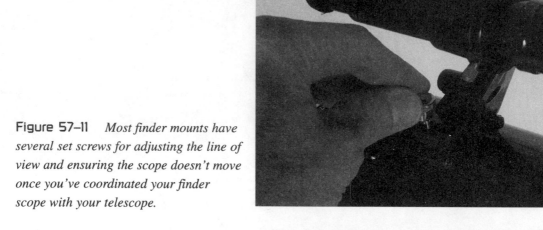

Figure 57–11 *Most finder mounts have several set screws for adjusting the line of view and ensuring the scope doesn't move once you've coordinated your finder scope with your telescope.*

Figure 57–12 *Ready for prime time. A properly aligned finder scope can mean the difference between a productive night's viewing and slewing around the sky with nary a chance of locating your desired object.*

Project 58. How to Assemble a Professional-Grade Telescope

What You Need:

- Astronomy Technologies AT66 60mm f/6 ED refractor

- William Optics UWAN 28mm 2-inch eyepiece

- AOK (William Optics) Easy Touch ALT-AZ mount (plus tripod)

Resources:

- Astronomy Technologies, Inc.— www.astronomytechnologies.com

- William Optics—www.williamoptics.com

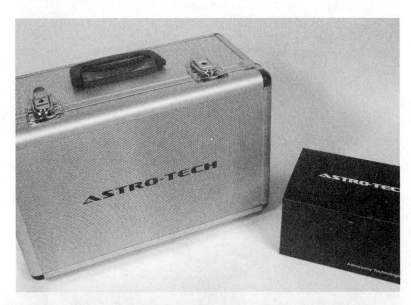

Figure 58–1 *Astronomy Technologies (or Astro-Tech) AT66ED High-Quality Compact Apochromatic Refractor telescope with SCT-Type 2-inch Diagonal.*

Figure 58–2 *This extra-low dispersion glass element 400mm f/6 telescope delivers professional-grade apochromatic performance.*

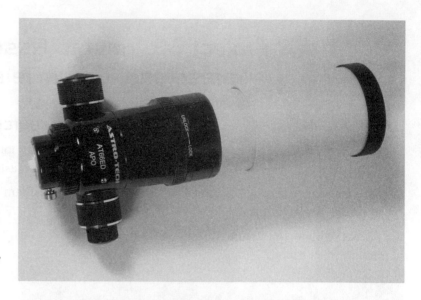

Figure 58–3 *The Astro-Tech AT66ED is finished with a durable paint and liquid-anodized high-gloss finish. It looks as beautiful as it works.*

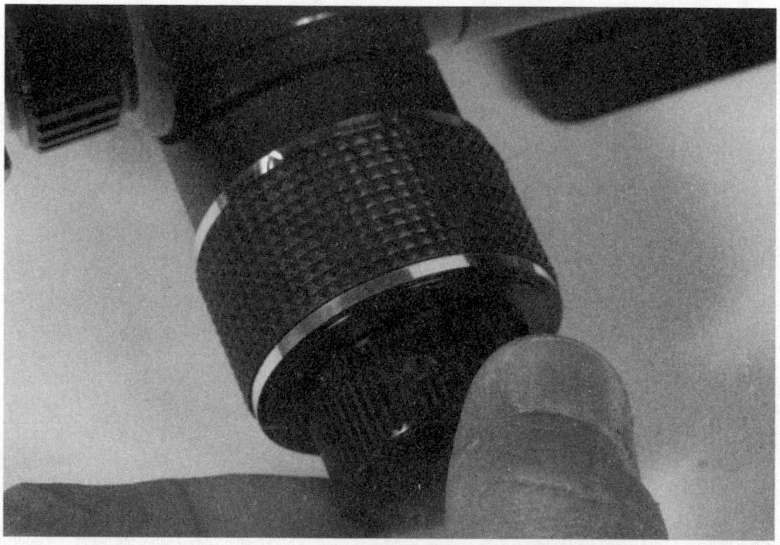

Figure 58–4 *One fine-focus knob is provided.*

Figure 58–5 *Lock the focusing knobs in place with a hex wrench.*

Figure 58–6 *A high-quality metal lens cap protects the multicoated lens.*

Figure 58–7 *A retractable lens shade can help to reduce lens glare.*

Figure 58–8 *Remove the 1.25-inch eyepiece holder for adding a diagonal.*

Figure 58–9 *This diagonal will accept either 1.25-inch or 2-inch eyepieces.*

Figure 58–10 *Ready for a 1.25-inch eyepiece.*

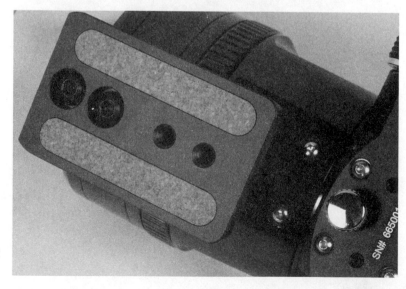

Figure 58–11 *The Astro-Tech AT66ED can be mounted either on a conventional photography tripod or on an astronomical L-bracket dovetail mount.*

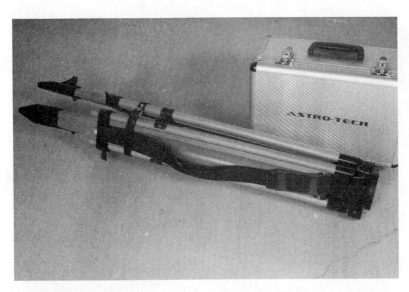

Figure 58–12 *William Optics wooden tripod is ideal for holding the Astro-Tech AT66ED refractor telescope.*

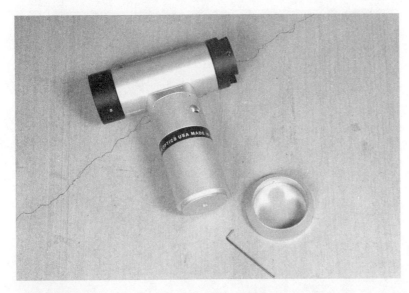

Figure 58–13 *William Optics Eazy Touch ALT-AZ mount. Actually, this mount is manufactured in conjunction with AOK in Switzerland.*

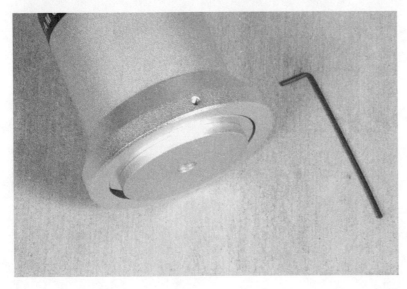

Figure 58–14 *Loosely attach the tripod mount collar.*

Figure 58–15 *Loosely thread the dovetail mount set screw into place.*

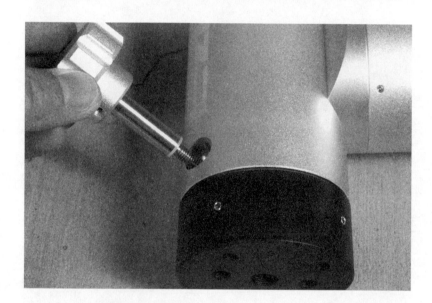

Figure 58–16 *Loosely thread the altitude adjustment set screw into place.*

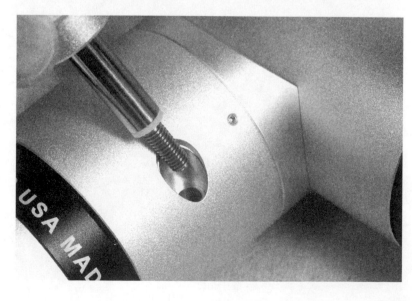

Figure 58–17 *Loosely thread the azimuth adjustment set screw into place.*

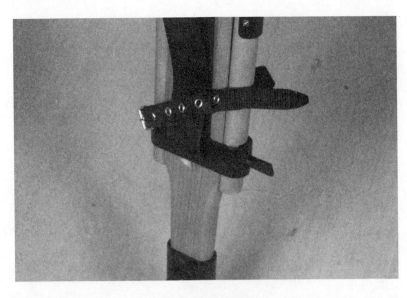

Figure 58–18   *Open the wooden tripod and wrap the retaining strap around one leg.*

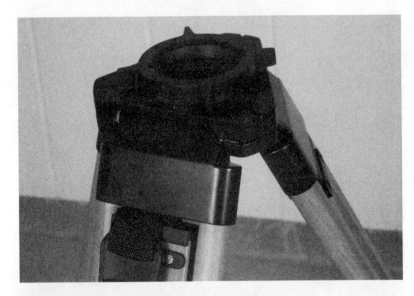

Figure 58–19   *Level the tripod.*

Figure 58–20   *Remove the tripod mount locking nut.*

Figure 58–21 *Carefully drop the tripod mount plate into place. Watch out for the thread grease.*

Figure 58–22 *Reattach the tripod mount-locking nut.*

Figure 58–23 *Tighten the locking nut.*

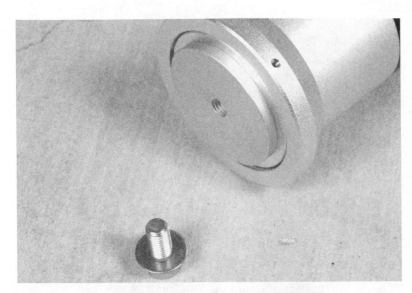

Figure 58–24 *Locate the mount attachment bolt and add a washer over the threads.*

Figure 58–25 *Carefully set the mount on top of the tripod mount plate.*

Figure 58–26 *Insert the mount attachment bolt.*

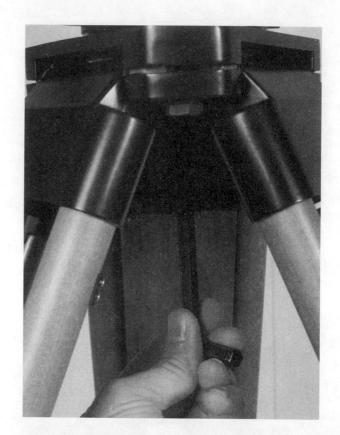

Figure 58–27 *Tighten the mount attachment bolt with a hex wrench.*

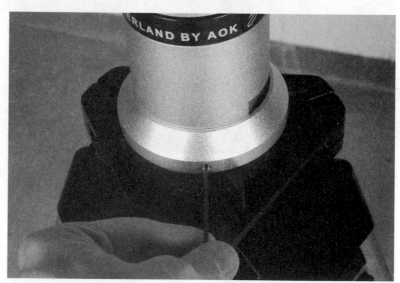

Figure 58–28 *Tighten the tripod mount collar.*

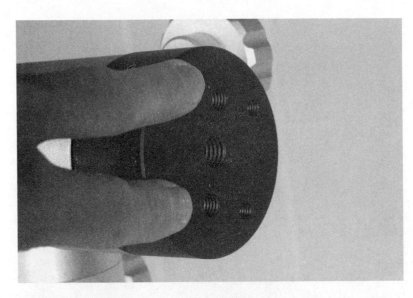

Figure 58–29 *This is where you mount the Vixen-style dovetail telescope mount plate.*

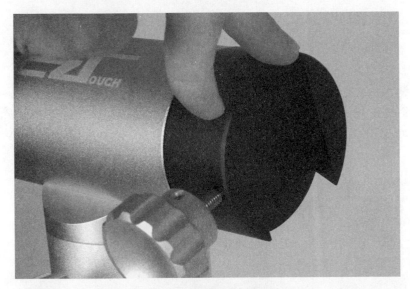

Figure 58–30 *This is where you can mount a dovetail bracket telescope, such as the Astro-Tech AT66ED.*

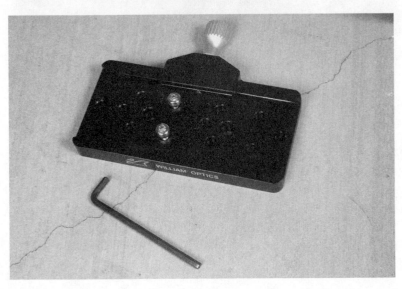

Figure 58–31 *Prepare the Vixen-style dovetail telescope mount plate.*

Figure 58–32 *Tighten the Vixen mount plate bolts with a hex wrench.*

Figure 58–33 *The Vixen-style dovetail telescope mount plate.*

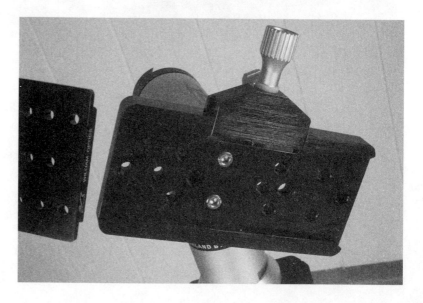

Figure 58–34 *A telescope receiving plate slides into the mount.*

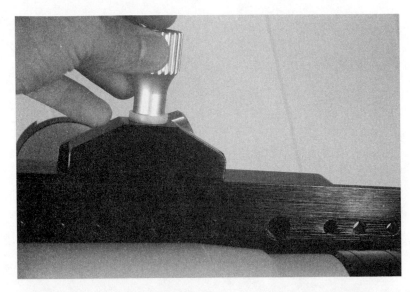

Figure 58–35 *And the plate is locked in place.*

Figure 58–36 *William Optics UWAN 82° 28mm 2-inch eyepiece.*

Figure 58–37 *This is a professional-grade eyepiece.*

Figure 58–38 *Mount the UWAN eyepiece in the Astro-Tech diagonal.*

Figure 58–39 *Mount the diagonal on the AT66ED refractor telescope.*

Figure 58–40 *This is a great eyepiece/telescope combination for viewing the moon and planets.*

Figure 58–41 *The Astro-Tech AT66ED includes a locking collar, enabling you to rotate the focuser to any desired viewing angle.*

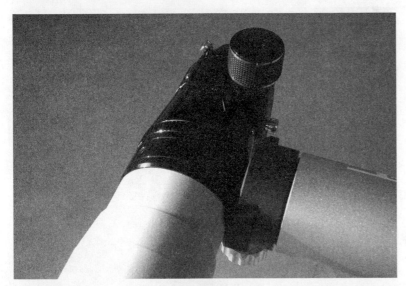

Figure 58–42 *Slip the AT66ED into the dovetail mount of the Eazy Touch ALT-AZ mount.*

Figure 58–43 *Either a diagonal or an eyepiece can be added to this telescope.*

Figure 58–44   *You can rotate the focuser, so the coarse focus knob is up.*

Figure 58–45   *This mount is able to accommodate two telescopes at the same time.*

Project 59. Seeing Solar Sights

What You Need:

- Zhumell Aurora 70

Resources:

- Zhumell, Inc.—www.zhumell.com

Figure 59–1 *Never view the sun without using an approved solar filter.*

Figure 59–2 *You can safely project an image of the sun onto a sheet of cardstock. Begin by removing the eyepiece of your telescope.*

Figure 59–3 *Add a solar shade. This sheet of cardstock will block extraneous sunlight from your projection sheet and will hide the finder scope from an accidental solar view.*

Figure 59–4 *Return the eyepiece.*

Figure 59–5 *Remove only the center portion of your telescope's lens cap.*

Figure 59–6 *Move the projection sheet until you get a sharp image of the sun.*

Project 60. How to Trail Stars

What You Need:

- Nikon FM10 SLR film camera (yes, *film*, not digital)
- Vivitar 500mm mirror lens

Resources:

- Nikon U.S.A.—www.nikonusa.com
- Vivitar—www.vivitar.com

Figure 60–1 *Yeah, remember this stuff? It's called photographic film. And it's a great medium for use in astrophotography.*

Figure 60-2 *You will also need one of these—a film SLR camera.*

Figure 60-3 *Nikon FM10. This is a great low-cost film SLR camera for testing the astrophotography waters.*

Figure 60-4 *Only one little piece of modern technology is inside the FM10— a light meter battery. If you plan on using your film camera as a dedicated astrophotography camera, you won't need this newfangled contraption, so you can remove the battery.*

Figure 60–5   *Save the battery for later, when you want to return the camera to more conventional and down-to-Earth photography.*

Figure 60–6   *Remove the lens.*

Figure 60–7   *Find a T-Mount camera adapter. I removed one from this Vivitar 500mm mirror lens.*

Figure 60–8 *You will be primarily using the B or Bulb shutter-speed setting for astrophotography.*

Figure 60–9 *A threaded shutter release button is needed for adding a cable release.*

Figure 60–10 *A cable release.*

Figure 60–11 *Thread the cable release into the shutter release button.*

Figure 60–12 *Ensure that you don't accidentally cross-thread the cable release.*

Figure 60–13 *Now you can remotely trigger the camera's shutter without touching the camera.*

Figure 60–14 *Make sure your cable release also has a locking screw for enabling l-o-n-g timed exposures.*

Project 61. Astrophotography

What You Need:

- Nikon FM10 SLR film camera
- Meade MECAVPSFE variable projection SLR camera adapter for 1.25-inch eyepieces
- Tele Vue TE2CA SLR camera adapter for 2-inch eyepieces
- T-Mount for Nikon

Resources:

- Meade Instruments Corporation— www.meade.com
- Nikon U.S.A.—www.nikonusa.com
- Tele Vue Optics, Inc.—www.televue.com

Figure 61–1 *Tele Vue® 2-inch Focuser to T-Ring Adapter (back) and Meade® Vari-Cam Adap-A-Size.*

Figure 61-2 *T-Mount camera adapters can be used for focusing images directly on a camera's film plane (digital or film).*

Figure 61-3 *Remove the lens.*

Figure 61-4 *Add the T-Mount to the camera adapter.*

Figure 61–5 *Mount the adapter on the camera and use the camera/adapter instead of an eyepiece.*

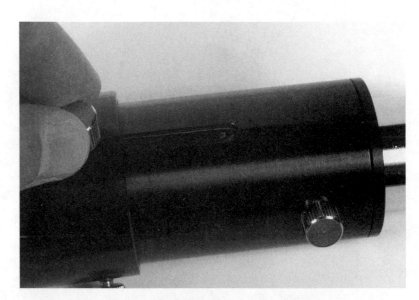

Figure 61–6 *The Meade Vari-Cam adapter can slide forward and backward for different magnifications.*

Figure 61–7 *The Tele Vue adapter behaves similarly to the Meade Vari-Cam adapter. Remove the camera's lens.*

Figure 61–8 *Add the T-Mount to the camera adapter.*

Figure 61–9 *Mount the adapter on the camera and use the camera/adapter in lieu of an eyepiece.*

Project 62. The Ideal Star Party Kit

What You Need:

- Great Red Spot Junior Red LED Shake Flashlight
- Magellan eXplorist 210
- Night Owl Optics Night Vision Night Scope

Resources:

- Great Red Spot Astronomy Products—www.greatredspot.com
- Magellan GPS—www.magellangps.com
- Night Owl Optics—www.nightowloptics.com or 1-800-444-5994

Great Red Spot Junior Red LED Shake Flashlight Physical Specifications:

- Light Output—Equivalent to a five AA battery flashlight (10,000 Lux)
- Light Beam—Circular beam with a diameter of 4.6ft or 1.4m at 5m distance and visible for 1 mile
- Light Duration—1–2 hours

- Energy System—Renewable magnetic charging system
- Switch—Non-contact luminescent switch glowing for days
- Safety—No spark, fireproof, and explosion-proof
- Unit Dimensions—6" (Length) by 1.5" (Diameter)
- Unit Weight—6.8 ounces
- Submersion—Waterproof, floatable, and operates to 410-ft or 125m water depth
- Corrosion/Shock—Corrosion-proof and survives a 6.6-ft drop
- Temperature—Functions between –240° F (–400° C) and +1400° F (+760° C)
- Lifespan—Operates for 6–10 years
- 20 seconds of shaking provides 20–30 minutes of white light
- 30 seconds of shaking provides 1–2 hours of white light
- Backed by a 6-year warranty

Figure 62–1 *The three essential devices for a star party field kit: Great Red Spot Junior Red LED Shake Flashlight (left), Magellan eXplorist 210 (center), and Night Owl Optics Night Vision Scope (right).*

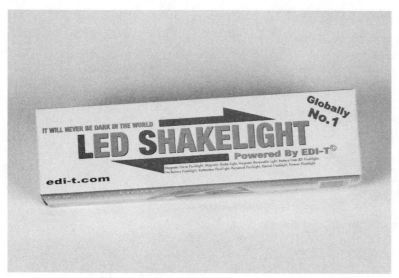

Figure 62–2 *Great Red Spot Junior Red LED Shake Flashlight.*

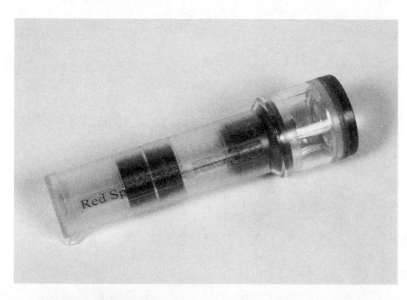

Figure 62–3 *A built-in red LED provides the ideal illumination without robbing you of your night vision.*

Figure 62–4 *A lens helps focus the LED light to a bright beam.*

Figure 62–5 *Power for the flashlight is provided by moving a magnet back and forth through a coil.*

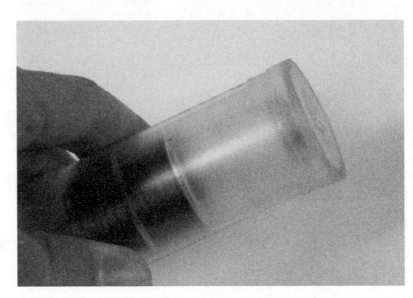

Figure 62–6 *Twenty seconds of shaking the flashlight can generated 20–30 minutes of usage.*

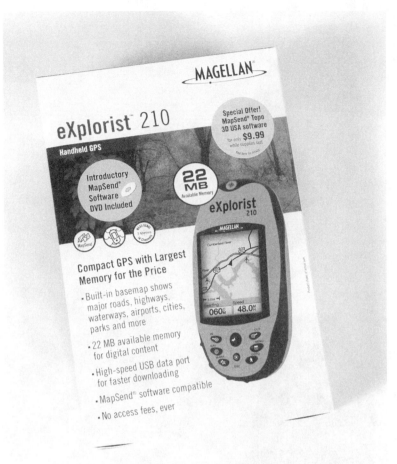

Figure 62–7   *Magellan eXplorist 210.*

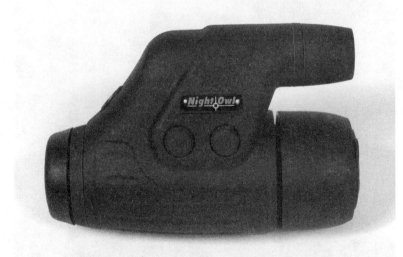

Figure 62–8   *Night Owl Optics Night Vision Scope.*

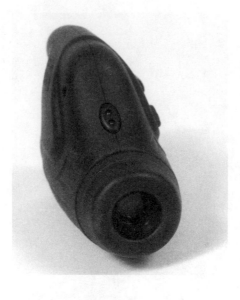

Figure 62–9 *This monocular form factor is ideal for checking your ground site for obstructions.*

Figure 62–10 *The lens cap also functions as a daylight shield for the sensitive night-vision sensor.*

Figure 62–11 *If you need some extra night illumination, a built-in IR beam can be used for "exposing" anything in view, up to 125 yards away.*

Project 63. Do-It-Yourself Deep-Sky Photography

What You Need:

- Nikon FM10 SLR film camera
- Zoom Nikkor 35–70mm f/3.5-4.8 lens
- Zhumell Aurora 70

Resources:

- Nikon U.S.A.—www.nikonusa.com
- Zhumell, Inc.—www.zhumell.com

Figure 63–1 *A regular film camera is ideal for making some deep-sky photographs.*

Figure 63–2 *Long exposures and a motor drive for your telescope are mandatory for deep-sky photography.*

Figure 63–3 *Use cable ties for mounting your camera on your telescope.*

Figure 63–4 *Loop a cable tie through your camera's case for providing a stable attachment platform.*

Figure 63–5 *If you'd like to photograph star trails, then mount your camera on a conventional photography tripod and center your field-of-view on Polaris.*

Figure 63–6 *When mounting your camera on your telescope for deep space photography, make sure the camera's field-of-view clears all obstructions on the telescope. This check is best performed during the daylight, before you begin to take your photographs.*

Figure 63–7 *Align the camera perfectly with the telescope tube. Then, the telescope's motor drive can be used for tracking your deep-sky object.*

What You Need:

- Digital camera; rangefinder or SLR
- Charged-coupled device (CCD) digital camera image sensor

Figure 64–1 *If you're willing to do some hacking, this disposable camera can be used as an Astrocam.*

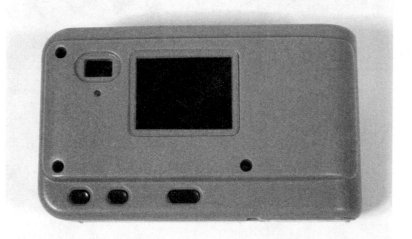

Figure 64–2 *The built-in LCD screen can be used for checking your astrophotography.*

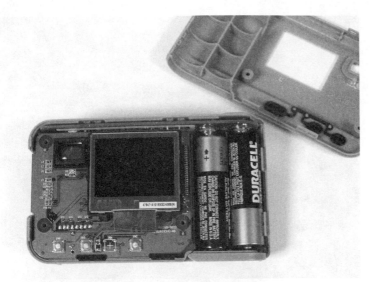

Figure 64-3   *Remove the case back panel.*

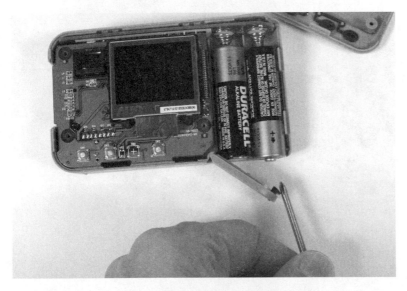

Figure 64-4   *Remove the batteries.*

Figure 64-5   *Carefully remove the camera's circuit board.*

Figure 64–6   *Don't touch this! This electrolytic capacitor is used for firing the electronic flash and it does retain a sufficiently painful charge—even when the camera has been turned off.*

Figure 64–7   *This is the lens/shutter assembly.*

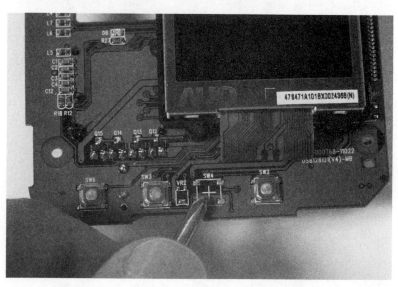

Figure 64–8   *Unfortunately, this orphaned switch pad doesn't provide you with any secret camera function. It is used for aborting an image-delete operation.*

Figure 64–9 *The CMOS sensor (black square along the left hand edge of circuit board).*

Figure 64–10 *The two wires are used for firing the shutter.*

Figure 64–11 *You can use John Maushammer's hack for enabling the camera's edge connector and download your images to your computer. See www.camerahacking.com for more information.*

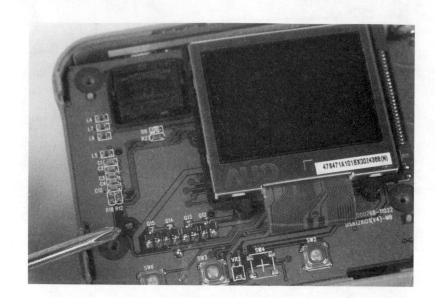

Figure 64–12 *The lens/shutter assembly is held in place by this screw.*

Figure 64–13 *And this screw.*

Figure 64–14 *This CCD sensor for a 3Com Webcam is a better image maker for astrophotography.*

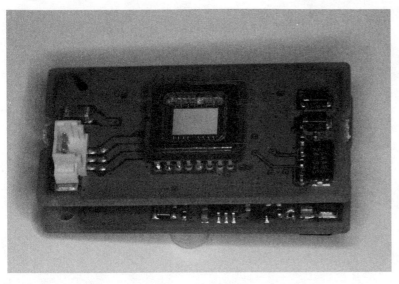

Figure 64–15 *You can buy CCD sensors, but they are impractical for DIY astrophotography.*

Figure 64–16 *This Sharp CCD sensor was removed from another webcam.*

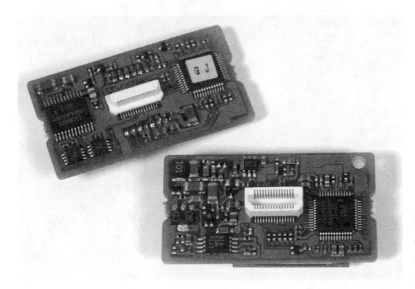

Figure 64–17 *Lacking adequate documentation, this CCD sensor board combination is also impractical for DIY astrophotography.*

Figure 64–18 *A discarded digital camera is a better starting point for building your own DIY Astrocam.*

Figure 64–19 *This old Konica Minolta DiMage Z2 is perfect for conversion into a CCD camera.*

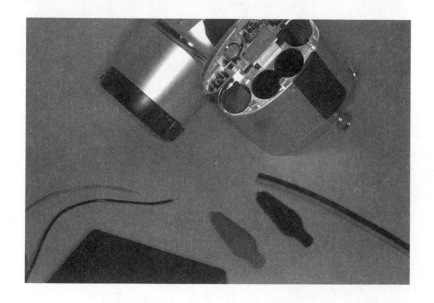

Figure 64–20 *Build an external battery pack.*

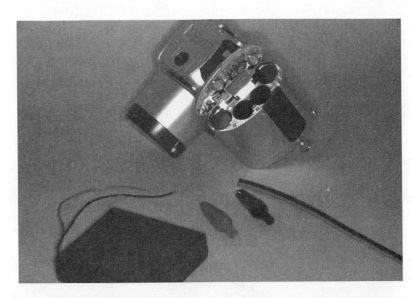

Figure 64–21 *The Z2 uses four AA batteries for power.*

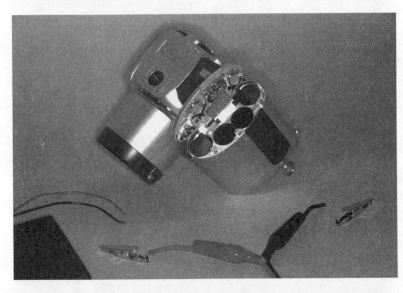

Figure 64–22 *The external battery pack is connected to the camera via two insulated alligator clips.*

Figure 64–23 *Insert the clips into the two outside battery holders.*

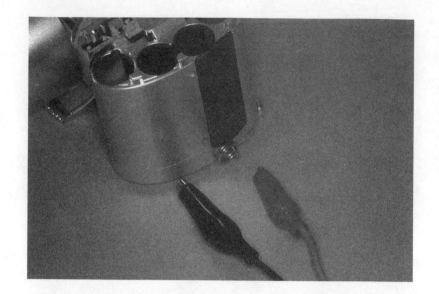

Figure 64–24   *Use a digital multimeter to verify which battery holder is positive (+) and which one is negative (–).*

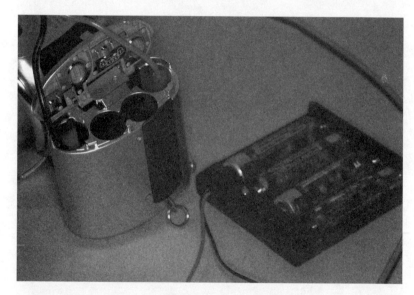

Figure 64–25   *Clip the external battery pack in place.*

Figure 64–26   *Test the camera for proper operation.*

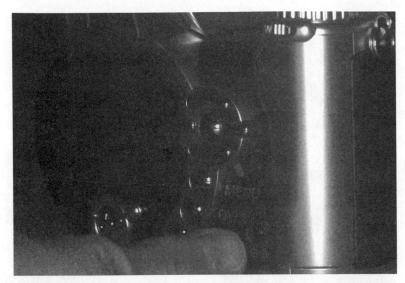

Figure 64–27 *Resist the temptation to use your external battery pack through the AC adapter port. The required voltage must be a constant and stable 6V. Any loss in power and the camera will turn itself off.*

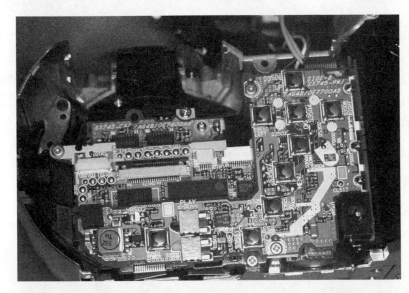

Figure 64–28 *Carefully remove all the screws holding the outside case together and open 'er up. Several ribbon cables and wire connectors are attached to the various components (for example, the LCD has three ribbon cables and one wire connector). Make sure you disconnect these cables and connectors before you separate the two case halves.*

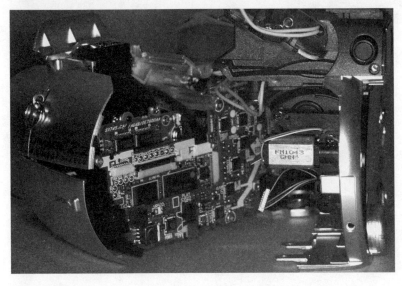

Figure 64–29 *The LCD (right) and the lens assembly (left).*

Figure 64–30 *Underneath the main circuit board is the CCD sensor board. The black plastic is insulation between the circuitry and the SD card slot located on the bottom of the main circuit board.*

Figure 64–31 *The SD card slot (top left).*

Figure 64–32 *The CCD sensor. An infrared (IR) filter is attached to the front of the sensor.*

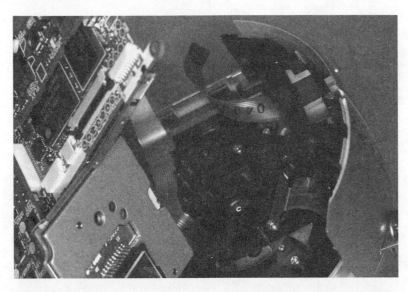

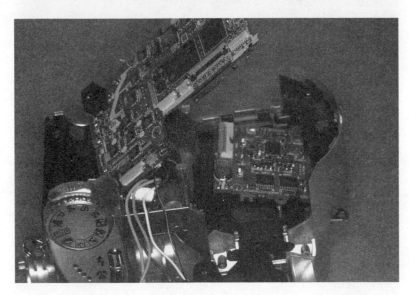

Figure 64–33 *The lens assembly (right). Remove this assembly.*

Figure 64–34 *Remove the IR filter from the face of the CCD sensor for increasing your camera's IR sensitivity.*

Figure 64–35 *Reinstall the CCD sensor circuit board.*

Figure 64–36 *The lens for the Z2. Most digital cameras need the lens for the proper execution of the camera's startup sequence. Basically, the camera "expects" to receive some feedback from the lens (for example, did the lens extend, is the shutter OK, is the aperture OK, and so forth). Without this feedback, the camera goes stupid and becomes a brick. So, remove the lens, but keep it attached to the camera.*

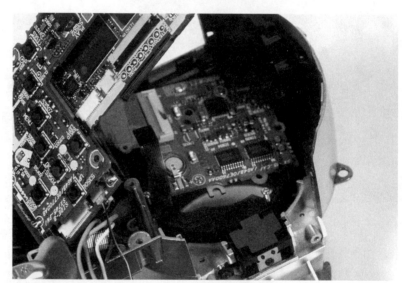

Figure 64–37 *As you return each component, reattach its ribbon cable and wire connector.*

Figure 64–38 *Ideally, the Thermoelectric Heat Pump Peltier Junction (the large white square in the center) should go here—right behind the CCD sensor. Space constraints in the Z2 mandated that the Peltier Junction be moved to the bottom of the camera.*

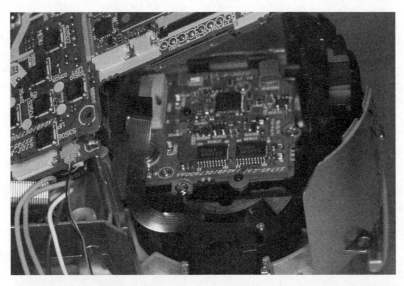

Figure 64–39 *The CCD sensor circuit board mounted to the back of the empty lens assembly.*

Figure 64–40 *Remove the tripod socket from the bottom of the camera. This will facilitate the cool air movement from the Peltier Junction.*

Figure 64–41 *All closed up, wired up, and ready for action.*

Figure 64–42 *The Peltier Junction can be activated with a separate switch.*

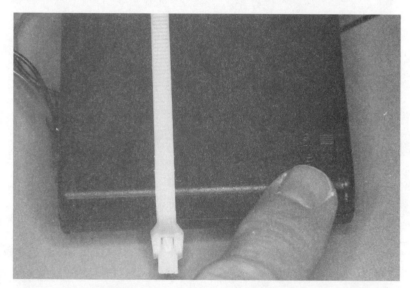

Figure 64–43 *Another switch is used for enabling the external battery pack.*

Figure 64–44 *Make sure all heat-sensitive items (such as these lens ribbon cables) stay away from the Peltier Junction heat sink.*

Figure 64–45 *Turn on the external battery pack, and then turn on the camera.*

Figure 64–46 *I dropped my lens assembly, but I retained the motors and switches that drive the lens, shutter, and aperture. To operate these switches, I have to short them with a clip during the camera startup sequence.*

Figure 64–47 *It ain't pretty, but it works.*

Figure 64–48 *The shutter release button must either be able to capture long, extended exposures or you can use the video recording mode.*

Figure 64–49 *If you need one, a discarded 35mm film canister makes a great 1.25-inch eyepiece adapter.*

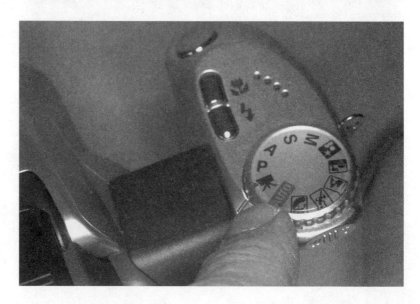

Figure 64–50 *The video recording mode is more practical for using this CCD camera for astrophotography.*

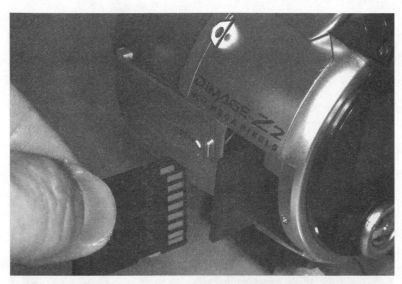

Figure 64-51 *Keep several large-capacity digital media cards formatted and ready for use. Shooting video eats up a lot of memory space.*

Figure 64-52 *Prior to heading to the telescope, make sure you shoot at least one dark field image (with the lens cap on) for checking the noise level (illuminated pixels) of your camera. Switch the Peltier Junction on and off in a couple of the shots and see if there is any improvement. If not, grab another discarded digital camera and try again.*

Project 65. Videotaping the Night Sky

What You Need:

- Canon Elura 2MC

Resources:

- Aiptek—www.aiptek.com
- Canon U.S.A.—www.usa.canon.com

Figure 65–1 *Canon Elura 2MC (left) and Aiptek IS-DV (right).*

Figure 65–2 *Hold your digital video camera up to the eyepiece for a painless video tour of the heavens.*

Figure 65–3 *You can check your video in the field. Cover the LCD with a piece of red cellophane during viewing. This will protect your night vision.*

Chapter Eight

Your First Job Is on Mars

In the annals of space exploration, our path to the stars has been paved by robots. Some staunch robot physicists might even argue that robots not only paved the way to outer space, but they also surveyed, graded, and maintained the small highways and byways we've carved into the great space around us.

While today's Martian rover robots grab most of the headlines, one of the earliest NASA robots was launched in the early 1960s. This robot was named Mariner 2. Later, in 1962, Mariner 2 made a flyby of Venus, and it transmitted temperature and atmospheric data back to Earth.

Mars robotic exploration is still a hot topic (see Figure 8-1). And, on March 17, 2006, NASA named John Callas as the project manager for NASA's Mars Exploration Rover missions, including the Rover named "Spirit." Callas is a scientist at NASA's Jet Propulsion Laboratory (JPL) in Pasadena, CA.

"It continues to be an exciting adventure with each day like a whole new mission," Callas said.

That modesty is the epitome of understatement. For example, one of Spirit's six wheels had stopped working. Dragging that wheel, Spirit had to sprint to a slope where it could catch enough rays to continue operating during the upcoming Martian winter. That period of minimum sunshine was more than 100 days away, but Spirit could only muster enough power for about one hour per day of driving on flat ground.

And you think your earthbound evil robot competitions are tough?

The best spot for Spirit's "snow bird" migration was the north-facing side of McCool

Figure 8–1   *A bird's-eye view self-portrait of NASA's Mars Exploration Rover Spirit. (Photography courtesy of NASA/JPL-Caltech/Cornell.)*

Hill, where it spent the southern-hemisphere winter tilted toward the sun just soakin' in the fun.

During Spirit's sprint toward the hill, it traveled approximately 120 meters (about 390 feet). Unfortunately, this robot's flat-out speed was approximately a whopping 12 meters (40 feet) per day.

Well, did Spirit make it? Yes, Spirit was able to wait out the Martian winter, and slowly in

September 2006, electrical power began to surge from its photovoltaic arrays. Robot life on Mars was ready to begin anew.

Resources:

- Mars Exploration Rover Mission— marsrovers.nasa.gov/home/

- NASA Robotics: The Robotics Alliance Project—robotics.nasa.gov

Project 66. Mars Driver's Ed 101

What You Need:

- JoinMax Digital Tech. RoboEXP Educational Kit 0600

Resources:

- JoinMax Digital Tech. Ltd.— www.robotplayer.com

- JoinMax RobotEXP—www.robotexp.com/

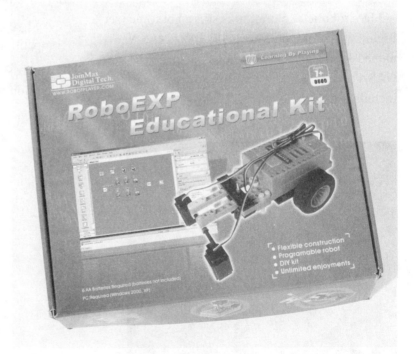

Figure 66–1 *JoinMax Digital Tech Ltd. RoboEXP Educational Kit 0600.*

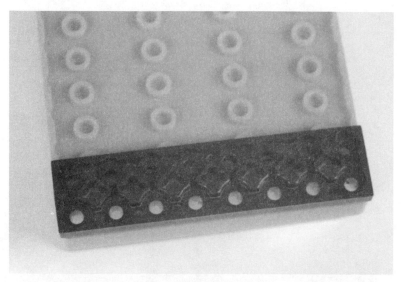

Figure 66–2 *Using easy-to-assemble plastic plates that snap together, the JoinMax 0600 kit is a great DIY robot design kit.*

Figure 66–3 *The kit comes with two drive motors.*

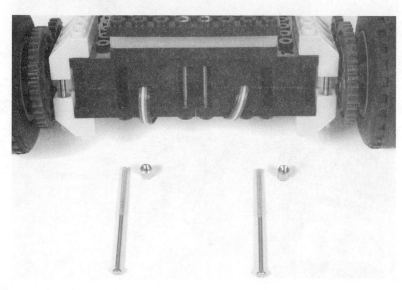

Figure 66–4 *All the necessary parts are included for building two gear-drive wheels.*

Figure 66–5 *Assembled plastic parts can be reinforced with bolts and nuts.*

Figure 66–6 *A programmable controller, the Super Robot Control Unit (RCU), controls your robot.*

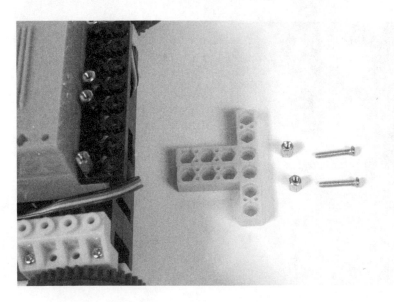

Figure 66–7 *Complex support structures can be used for holding sensors.*

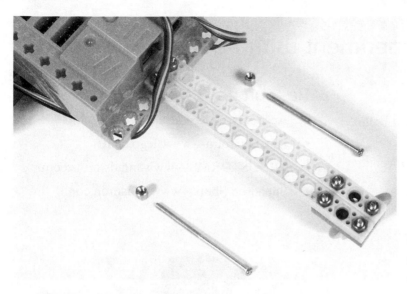

Figure 66-8 *A variety of fasteners can be used for ensuring that all components are held tightly together.*

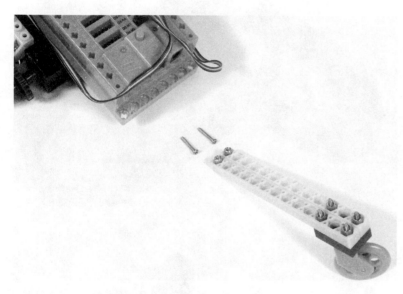

Figure 66-9 *This tail wheel can be mounted directly on the Super RCU.*

Project 67. Experiment with Feet of Tread

What You Need:

- Academy Leclerc French Army Main Battle Tank (MBT) model kit (No. 13001)
- LEGO® MINDSTORMS® Robotic Invention System™ 2.0

Resources:

- Academy—www.academyhobby.com
- LEGO Group—www.lego.com
- MINDSTORMS—www.mindstorms.com
- Squadron Shop—www.squadron.com

Figure 67–1 *This is the perfect motivator for any Mars Rover bot—tank treads.*

Figure 67–2 *A tank tread system built from the original LEGO® MINDSTORMS® Robotic Invention System (RIS).*

Figure 67-3 *Model tank kits, like this Academy LeClerc French Army Main Battle Tank hobby model kit, are terrific sources for tank treads.*

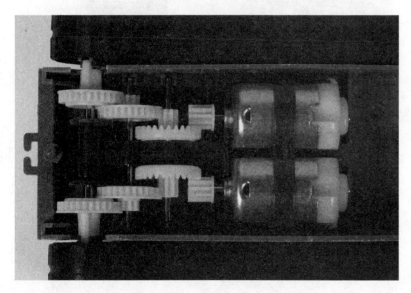

Figure 67-4 *The Academy LeClerc kit includes a preassembled motor gearbox with tank tread attachment gears.*

What You Need:

*   JoinMax Digital Tech. RoboEXP Educational Kit 0600

Resources:

*   JoinMax Digital Tech. Ltd.— www.robotplayer.com

*   JoinMax RobotEXP—www.robotexp.com/

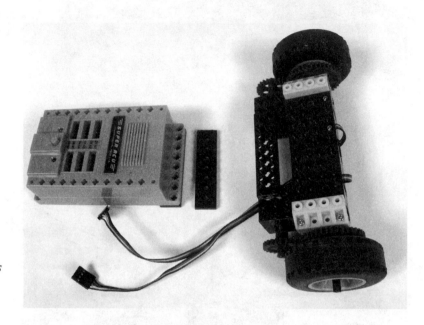

Figure 68–1 *Two geared wheel drives for the JoinMax 0600 DIY robot design kit.*

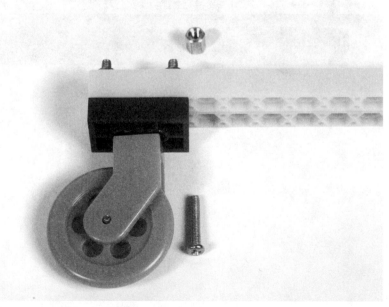

Figure 68–2 *A nonpowered tail wheel can be added to your JoinMax robots.*

Project 69. Build a Solar Power Array

What You Need:

- Eight BPW33 diodes
- Four 1000µF electrolytic capacitors
- Two 165-ohm resistors
- Two LEDs
- One 1N914 diode

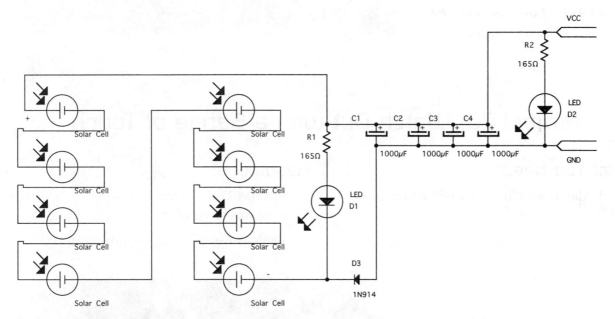

Figure 69–1   *Use this solar-powered generator circuit as the basis for your own solar systems.*

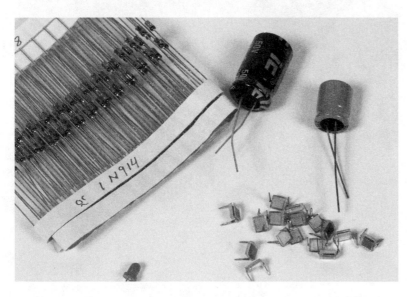

Figure 69–2   *Just a handful of discrete analog components are needed for building your own solar power generator. The BPW33 photodiodes (lower right) also generate electricity when exposed to light.*

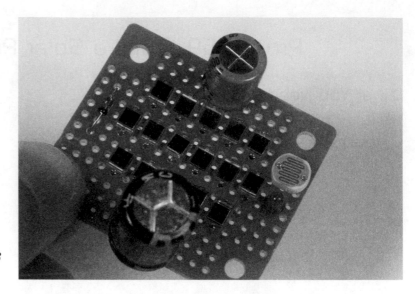

Figure 69–3 *Experiment with BPW33 photodiodes as a power supply.*

Project 70. Watch Out with a Sense of Touch

What You Need:

- JoinMax Digital Tech. RoboEXP Educational Kit 0600

Resources:

- JoinMax Digital Tech. Ltd.— www.robotplayer.com

- JoinMax RobotEXP—www.robotexp.com/

Figure 70–1 *A JoinMax 0600 RoboEXP Educational Kit robot touch sensor.*

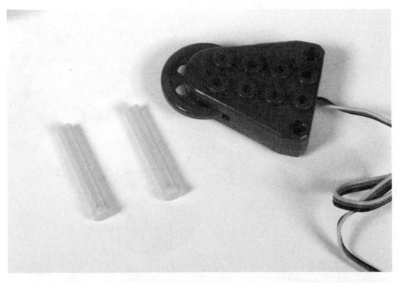

Figure 70–2 *A touch sensor can be added to any design with just a couple of mounting pins.*

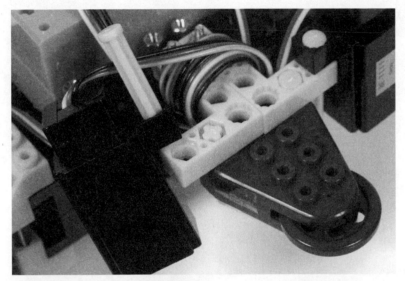

Figure 70–3 *A touch sensor mounted alongside two light sensors.*

Figure 70–4 *The touch sensor will help to protect the delicate light sensors for a collision.*

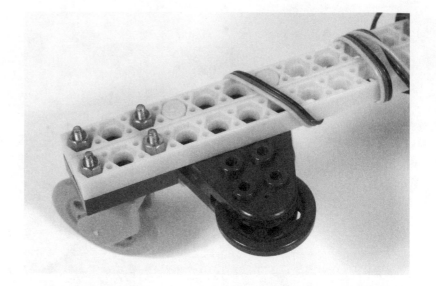

Figure 70-5 *Use a touch sensor to detect the swing of your tail wheel.*

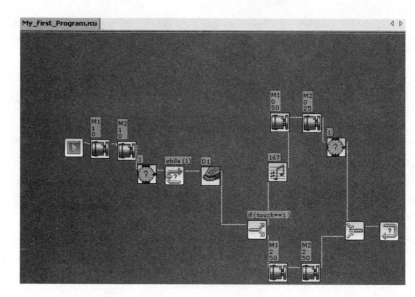

Figure 70-6 *You can integrate the control of the motors and your sensors with the easy-to-use programming environment included with the JoinMax DIY kit.*

Project 71. Move Toward the Light

What You Need:

- JoinMax Digital Tech. RoboEXP Educational Kit 0600

Resources:

- JoinMax Digital Tech. Ltd.— www.robotplayer.com
- JoinMax RobotEXP—www.robotexp.com/

Figure 71–1 *Several nifty light sensors are included with the 0600 JoinMax robot kit. These gray sensors can be used for following a line or checking for edges and cliffs. Yikes!*

Figure 71–2 *The JoinMax kit also includes a photoresistor light sensor (left) and an LED module (right).*

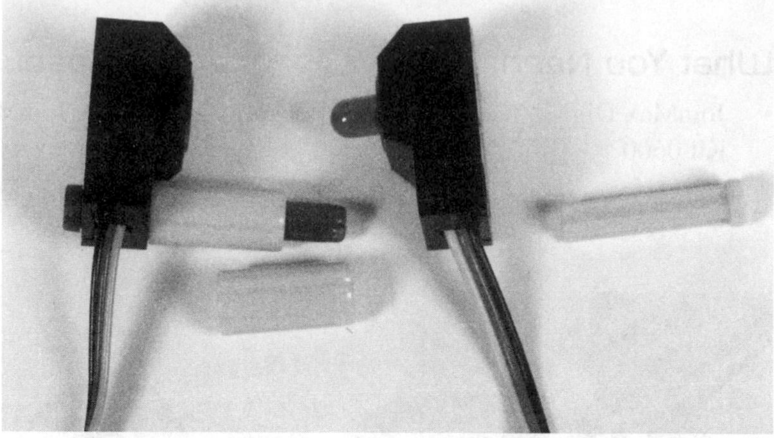

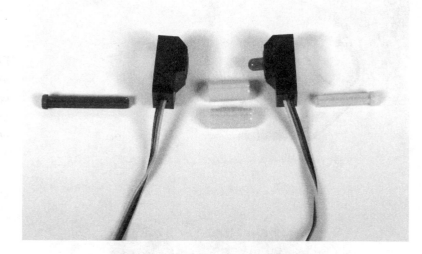

Figure 71–3 *A crude light-wave communication system can be built from the photoresistor light sensor and the LED module.*

Figure 71–4 *Mount the two sensors, facing each other.*

Figure 71–5 *Ensure that both sensors are extremely close to each other.*

Figure 71–6 *Mount this communication assembly on the underside of your robot. This placement will ensure that stray light doesn't affect your light wave communications.*

Project 72. Record a Martian Scream with a Sound Sensor

What You Need:

- JoinMax Digital Tech. RoboEXP Educational Kit 0600

Resources:

- JoinMax Digital Tech. Ltd.— www.robotplayer.com
- JoinMax RobotEXP—www.robotexp.com/

Figure 72–1 *This voice sensor in the JoinMax RoboEXP Educational Kit can be used for controlling your robot via sound/noise.*

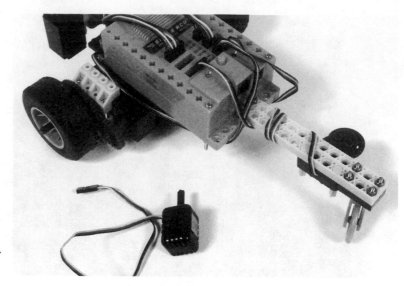

Figure 72-2 *Mount the sensor on your robot.*

Figure 72-3 *Route the connection wire from the voice sensor to one of the analog ports on the Super RCU.*

Project 73. Make Your Own Classy Chassis

What You Need:

- JoinMax Digital Tech. RoboEXP Educational Kit 0600
- LEGO MINDSTORMS NXT

Resources:

- JoinMax Digital Tech. Ltd.—www.robotplayer.com
- JoinMax RobotEXP—www.robotexp.com/
- LEGO Group—www.lego.com
- MINDSTORMS—www.mindstorms.com

Figure 73–1 *Ready to prowl any planet.*

Figure 73–2 *The JoinMax 0600 DIY kit can build a powerful rover robot bristling with sensors.*

Figure 73–3   *You program your robot with your PC computer, and then download the program into the Super RCU.*

Figure 73–4   *This LEGO MINDSTORMS NXT brick and ultrasonic sensor has been combined with a tank-tread propulsion system snatched from an Academy model tank kit. The connectors have been removed for clarity..*

Project 74. Track Your Movements

What You Need:

- JoinMax Digital Tech. RoboEXP Educational Kit 0600

Resources:

- JoinMax Digital Tech. Ltd.— www.robotplayer.com

- JoinMax RobotEXP—www.robotexp.com/

Figure 74–1 *Geared motors can be used for driving wheels faster or slower and with greater precision.*

Figure 74–2 *Exotic sensor combinations like this light-wave communication system (see Project 71) can be added to your rover robot.*

Figure 74–3 *Powerful data control systems like the Super RCU can enable your robot to master any terrain.*

Project 75. Give Your Rover Some Eyes

What You Need:

- JoinMax Digital Tech. RoboEXP Educational Kit 0600

Resources:

- JoinMax Digital Tech. Ltd.— www.robotplayer.com
- JoinMax RobotEXP—www.robotexp.com/

Figure 75–1 *These twin gray sensors can be used for detecting and following lines as the robot drives around.*

Project 76. Add Night Vision to Rover's Eyes

What You Need:

- 203CA wireless video camera
- JoinMax Digital Tech. RoboEXP Educational Kit 0600

Resources:

- JoinMax Digital Tech. Ltd.— www.robotplayer.com
- JoinMax RobotEXP—www.robotexp.com/

Figure 76–1 *A miniature battery-powered video camera can be added to your robot for remote monitoring.*

Project 77. Control Rover's Actions with a Programmable Brain

What You Need:

- JoinMax Digital Tech. RoboEXP Educational Kit 0600

Resources:

- JoinMax Digital Tech. Ltd.— www.robotplayer.com

- JoinMax RobotEXP—www.robotexp.com/

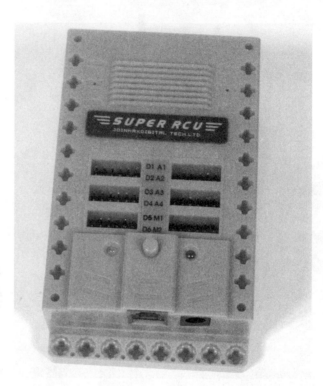

Figure 77–1   *The JoinMax Super RCU is a powerful robot controller.*

Figure 77–2   *You connect the Super RCU to your PC via the phone jack (bottom). This connection enables two-way communication with either your computer or another Super RCU.*

Figure 77–3 *An Atmel ATmega16 microcontroller drives the Super RCU.*

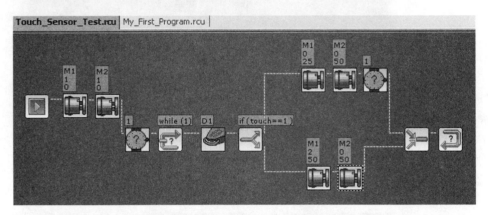

Figure 77–4 *Quick-and-easy programs can be written for the Super RCU with an intuitive, icon-based, drag-and-drop programming environment.*

Project 78. Monitor Your Sensors

What You Need:

- JoinMax Digital Tech. RoboEXP Educational Kit 0600

Resources:

- JoinMax Digital Tech. Ltd.—www.robotplayer.com

- JoinMax RobotEXP—www.robotexp.com/

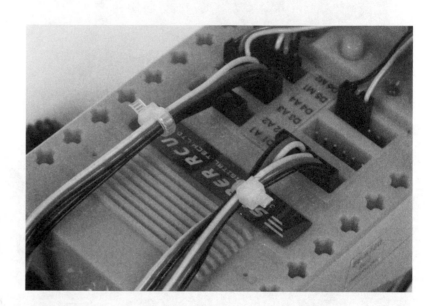

Figure 78–1 *Sensors and motors are plugged directly into the Super RCU. There are analog ports (A1–A4), digital ports (D1–D6), and motor ports (M1–M2).*

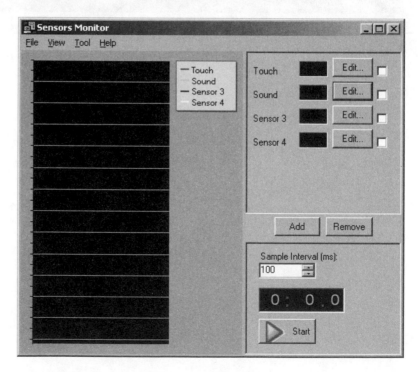

Figure 78–2 *A special Sensors Monitor program enables you to directly monitor all sensor readings from your robot on your PC.*

Project 79. Motivate Your Motors

What You Need:

- JoinMax Digital Tech. RoboEXP Educational Kit 0600

Resources:

- JoinMax Digital Tech. Ltd.— www.robotplayer.com
- JoinMax RobotEXP—www.robotexp.com/

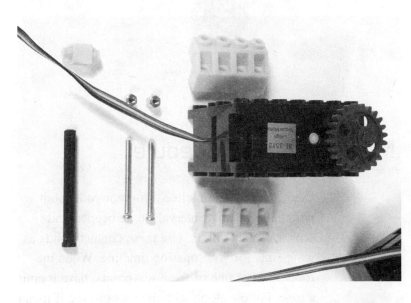

Figure 79–1 *The high-torque motors in the JoinMax 0600 DIY robot kit can be used for direct drive movement or configured to deliver a geared propulsion drive.*

Figure 79–2 *Try various gear combinations to get the ideal robot performance.*

Figure 79–3 *You can nest the two motors together for a single integrated drive unit that can be mounted directly underneath the Super RCU.*

Project 80. Maintain Your Schedule

What You Need:

- JoinMax Digital Tech. RoboEXP Educational Kit 0600

Resources:

- JoinMax Digital Tech. Ltd.— www.robotplayer.com

- JoinMax RobotEXP—www.robotexp.com/

Create a mission timeline. Program your robot with built-in timer intervals, tone beeper, and sensor contact points. Use these contact points as waypoints for your mission timeline. When the robot reaches one of these waypoints, have it emit a tone. Follow along with the robot to see if it can keep a schedule.

Project 81. Mission: Mars (OK, Your Backyard)

What You Need:

- JoinMax Digital Tech. RoboEXP Educational Kit 0600

Resources:

- JoinMax Digital Tech. Ltd.— www.robotplayer.com

- JoinMax RobotEXP—www.robotexp.com/

When you've debugged your robot design indoors, it's time to take that bot to Mars. If you have trouble convincing NASA about the merits of sending your robot design to Mars, transform your backyard into a Martian playground. A derelict sandbox makes a great site for creating your own tiny slice of Mars. Just add some rocks and turn your robot loose in this new sandy environment. Probably the first thing you'll notice is the sand has ruined your motors. Yup, if you're going to rove in the sand, then you'd better keep the sand out of your robot's internals.

Project 82. Learn How to Maintain Rover's Power Supply

What You Need:

- JoinMax Digital Tech. RobotEXP Educational Kit 0600
- Solar Power Array (see Project 69)

Resources:

- JoinMax Digital Tech. Ltd.— www.robotplayer.com
- JoinMax RobotExp—www.robotexp.com/

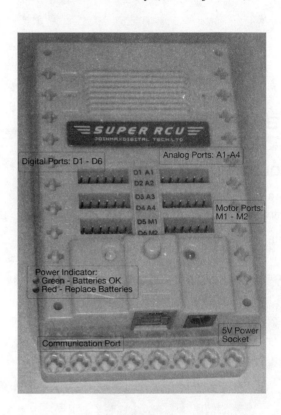

Project 83. Program Rover to Navigate Mars

What You Need:

- JoinMax Digital Tech. RoboEXP Educational Kit 0600

Resources:

- JoinMax Digital Tech. Ltd.— www.robotplayer.com
- JoinMax RobotExp—www.robotexp.com/

When your robot design has been properly shielded from sand, light, dust, and rocks (e.g., try Performix® Plasti-Dip® flexible DIY rubber coating material), then it's time to see if you can drive around in this harsh environment. Program your sensors (touch and light) to avoid obstacles and keep your rover safe.

Project 84. Record Mission Data

What You Need:

- JoinMax Digital Tech. RoboEXP Educational Kit 0600

Resources:

- JoinMax Digital Tech. Ltd.—
www.robotplayer.com

- JoinMax RobotEXP—www.robotexp.com/

During your Martian mission, set up a log book for recording waypoints, obstacles, program glitches, and rover video transmissions.

Project 85. Recover Rover and Build Rover II

What You Need:

- JoinMax Digital Tech. RoboEXP Educational Kit 0600

Resources:

- JoinMax Digital Tech. Ltd.—
www.robotplayer.com

- JoinMax RobotEXP—www.robotexp.com/

Figure 85–1 *You can design a wide variety of robots with the JoinMax RoboEXP Educational Kit 0600.*

Chapter Nine

Build Your Own ISS
in Your Backyard

Remember those wild designs for space stations back in the early 1960s? Invariably, each of these "artist conceptions" featured some sort of massive spinning wheel contraption with gigantic tubular spokes that traveled between the "hub" and the outer wheel.

Today's International Space Station (ISS) couldn't be more different (see Figure 9-1). Shaped like an Erector® set project gone awry, the ISS is the home for many strange and exciting

experiments. And none could be stranger than the building and launching of a manmade satellite constructed from a discarded Russian spacesuit.

Dubbed "SuitSat" (Spacesuit + Satellite), this outdated Russian Orlon® spacesuit was equipped with three batteries, a radio transmitter, and an array of sensors for reading temperature and battery strength. The plan called for SuitSat to circle the Earth and relay its status to ground stations monitoring its telemetry.

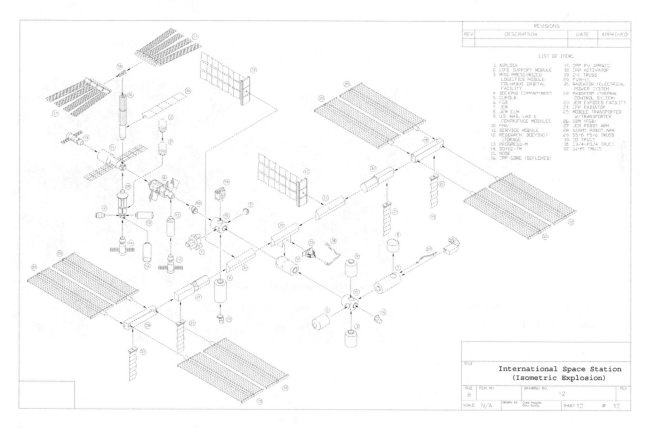

Figure 9-1 *Exploded view of International Space Station (ISS). (Drawing courtesy of NASA.)*

On February 3, 2006, at 11:27 P.M. EST, SuitSat-1 was launched into orbit during a 5-hour, 43-minute extravehicular activity (EVA) by ISS Expedition 12 Commander Bill McArthur and Flight Engineer Valery Tokarev. Launched into a "retrograde" orbit (an orbit that slows down and drops below the ISS), SuitSat was placed into orbit at the beginning of the second spacewalk of Expedition 12.

Believe it or not, according to members of the Connecticut-based American Radio Relay League (ARRL), this satellite actually worked. An ARRL member, Allen Pitts, reported on February 5, 2006, that his group had, indeed, received "weak, cold, and really hard to copy" transmissions from SuitSat.

Now if only an evil genius had a space station, what sorts of strange experiments could be performed?

Resources:

- An Empty Spacesuit Becomes an Orbital Experiment (NASA)— www.nasa.gov/mission_ pages/station/expeditions/expedition12/26jan_ suitsat.html

- SuitSat-1 RS0RS—www.amsat.org/amsat- new/ articles/BauerSuitsat/index.php

Project 86. How to Build a Geodesic Dome

What You Need:

- PVC tubing
- PVC connectors

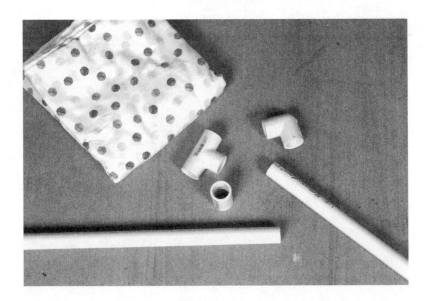

Figure 86–1 *You can fabricate a domed dwelling from conventional PVC tubing.*

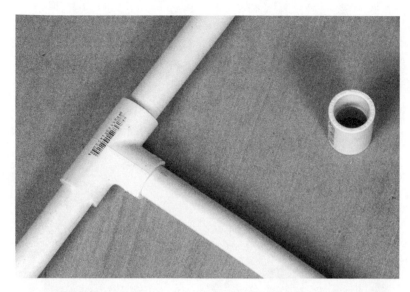

Figure 86–2 *Use T connectors for changing direction.*

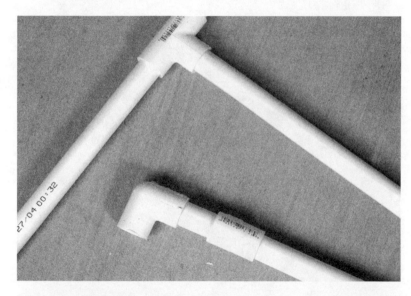

Figure 86–3 *Small extensions can be made from leftover pieces of tubing.*

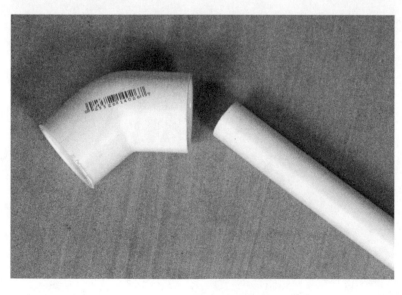

Figure 86–4 *A 45° angle connector is great for adding curves to your dome. This ½-inch tubing won't fit inside the 1-inch connector without an adapter.*

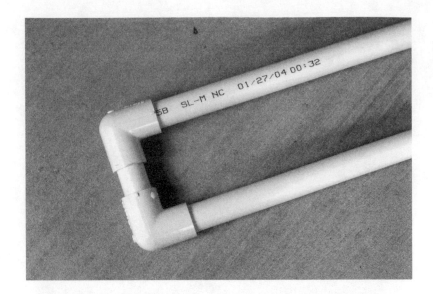

Figure 86–5 *A small spacer can tightly turn a corner with two 90° angle connectors.*

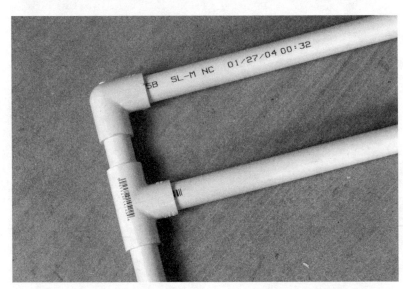

Figure 86–6 *Combine a 90° connector with a T connector for extending a wall side.*

Figure 86–7 *Even a small dwelling will require a lot of tubing and connectors.*

Figure 86–8   *If you can't design a suitable dome, try a more conventional dwelling shape.*

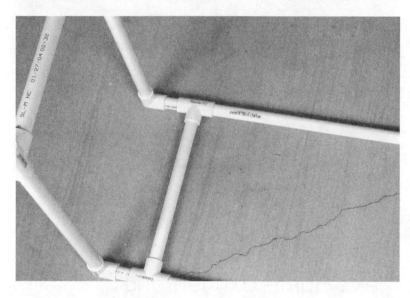

Figure 86–9   *In a conventional dwelling shape, you can PVC elbow connectors for building corners and raising walls.*

Project 87. Skin That Sucker:
Add an Exoskeleton to Your Spaceship

What You Need:

- Fabric
- Clear vinyl

Fabric is a common material used for covering domes. Glass, concrete, steel, and plastic are other materials that have been used for "skinning" geodesic domes.

Project 88. Build Your Own Solar Power Arrays

What You Need:

- 12-volt solar panel
- 12-volt deep cycle lead-acid battery
- DC voltage controller
- 12-volt DC meter
- DC-AC inverter
- Connection wWire

One of the best methods of generating power is with photovoltaic panels. Commonly called solar panels, these semiconductors are able to expel electricity when exposed to light. While this energy production output is terrific, the even better contribution from photovoltaic panels is that they don't produce any toxic exhaust waste. So, add a couple of these panels to your dome and live off the grid.

Project 89. Make Some Oxygen and
Get Rid of Carbon Dioxide

What You Need:

- Aquarium or water tank
- Aquatic plants
- Pump
- Tubing

Granted, you don't really have to worry about generating oxygen inside your terrestrial dome, but if you lived in a more austere environment, oxygen production would be your first priority. Aquatic plants can be helpful in meeting this requirement.

Project 90. What's for Dinner?

What You Need:

- Aquaculture
- Aquarium
- Hydroponics garden

Along with your aquatic plants, add some animals. This balanced ecosystem will also provide you with a source of protein. Yum, I can almost smell the boiled crawfish.

Project 91. Learn How to Distill Water

What You Need:

- Cardboard
- Plastic wrap
- Tape
- Weight

If you intend to survive inside your dome, then you'd better learn how to make your own water. Prepare your water distillation system by digging a small hole in the ground. Drop a cup into this hole (for gathering your water) and stretch some plastic wrap over this hole in the ground. Carefully drop a small weight onto the plastic, so it hangs over the cup. Now, sit back and catch some rays—solar rays, that is. As the sun heats the hole, water should condense on the plastic and drip into the cup. Cheers.

What You Need:

- Dome (see Project 86)
- Fan

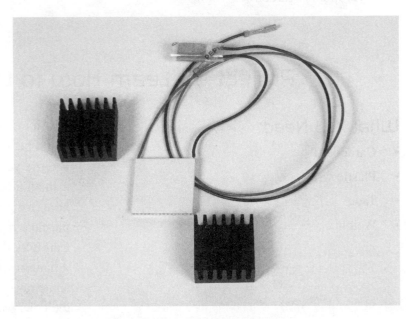

Figure 92–1   *You build a powerful cooling system from a Thermoelectric Heat Pump Peltier Junction. This small 30mm ceramic square (the white square in the center) just needs a heat sink and a power source (a 9V battery) to begin cooling.*

Figure 92–2   *You can find great cooling systems inside older PC computers, too. This fan assembly was removed from an Intel Pentium 4 microprocessor.*

Project 93. Put Waste in its Place

What You Need:

- Composting toilet

Waste management is no joking matter. Finding a clean, energy-efficient method for removing waste from your dome will be one of your most expensive challenges. Special composting toilets "recycle" waste. There is no need for either plumbing or power with these types of toilets.

Project 94. Keep Tabs on Your Mission: Monitor Yourself

What You Need:

- Polar S725 cycling heart rate monitor

Resources:

- Polar Heart Rate Monitors—
www.polarusa.com

If you're a professional athlete or an avid fitness nut, then you probably already own a heart rate monitor (HRM). An HRM is a wristwatch-like device that can measure your heart rate and track your exercise heart-rate limits throughout your entire daily exercise program. In the world of HRMs, Polar® Electro reigns supreme. Models like S725 are optimized for a specific type of exercise—biking. While the S725 excels at tracking your cycling regimen, it can also be used for tracking your daily health, wellness, and fitness.

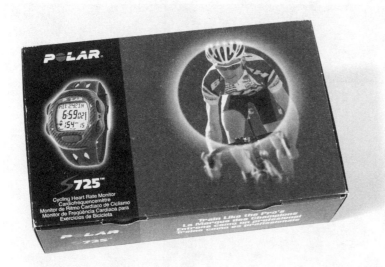

Figure 94–1 *Polar S725 cycling heart rate monitor (HRM).*

Figure 94–2 *While it looks just like a regular wristwatch, this Polar HRM is an essential link for monitoring your daily fitness.*

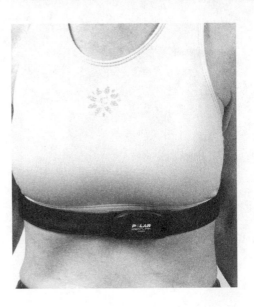

Figure 94–3 *Just strap it on your wrist.*

Figure 94–4 *Your body connection is also a wireless transmitter that monitors your heart rate during rest and while exercising.*

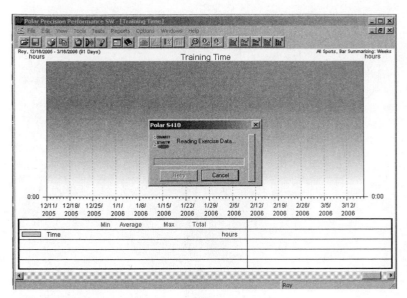

Figure 94–5 *Upload your S725 data to your PC via the watch's built-in infrared (IR) link.*

Project 95. Set a Record: Spend 439 Days Inside Your Space Station

What You Need:

- A whole lot of guts

Resources:

- Astronaut Bio: Shannon W. Lucid (188 days)—www.jsc.nasa.gov/Bios/htmlbios/lucid.html

- Biographies of USSR/CIS Cosmonauts Polyakov, Valeri (437 days from 1994–1995 aboard Soyuz TM-18)—www.spacefacts.de/bios/cosmonauts/ english/polyakov_valeri.htm

- NASA Astronaut Breaks U.S. Space Endurance Record (375 cumulative days in space)—www.nasa.gov/home/hqnews/2003/dec/HQ_03400_Foale_record.html

- NASA Astrobiology—astrobiology.arc.nasa.gov

- Reiter Breaks European Space Endurance Record (209 days in orbit)—www.esa.int/esaCP/SEME93JZBQE_index_0.html

Chapter Ten

You Mean the Big Dipper Isn't an Ice Cream Cone?

This should be a simple question to answer: Where is the oldest planetarium located? Right? Wrong.

If you qualify your question a little more with "Where was the oldest domed professional projector-equipped planetarium located?", then you should get the answer: the Carl Zeiss optical factory in Germany circa 1922. Thought to be the first professional planetarium projector, there is even more interest in the structure that housed this device.

Dr. Walter Bauersfeld was the inventor of this projector, but before he could adequately demonstrate its incredible capabilities, he also needed to design a "theater" for projecting his amazing light show. Bauersfeld settled on a dome design. And, therein, comes another interesting wrinkle in history. This structure was a dome fabricated from a wire mesh covered with a thin layer of concrete. Bearing a striking resemblance to a geodesic dome, Bauersfeld's dome predates R. Buckminster Fuller's geodesic dome patent by some 15 years (U.S. Patent 2,101,057, patented December 7, 1937).

Anyway, back to the Carl Zeiss planetarium.

About one year later, in the summer of 1923, Bauersfeld unveiled his night sky projection system to the public. The reviews were strongly supportive and the Carl Zeiss optical factory suddenly found itself in the planetarium projector business. Orders were welcome in depression-era Germany and 25 Zeiss projectors were built and shipped around the world prior to the start of the Second World War.

One of these Zeiss projectors found its way to the U.S. In 1930, the Adler Planetarium in Chicago, IL, became the first Zeiss-operated planetarium in the United States.

Now you can have the wonders of the night sky inside your own home. All it takes is a little evil this and a little evil that, and you'll soon be able to transform any room into a 360-degree recreation of the night's sky.

Resources:

- Celestia—www.shatters.net/celestia
- Stellarium—www.stellarium.org

Project 96. Plan a Trip to the Moon, Alice

What You Need:

- Uncle Milton Moon in My Room

Resources:

- Uncle Milton Toys—www.unclemilton.com

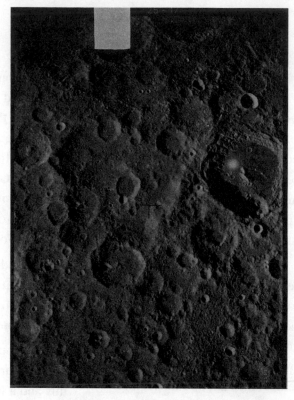

Figure 96–1 *On October 20, 1970, the Soviet Union launched a lunar probe named Zond 8 from the Baikonur Cosmodrome. During its seven-day mission, Zond 8 conducted circumlunar space investigations, as well as transmitting images of both Earth and the Moon. Zond 8 returned to Earth on October 27, 1970, with a recovery in the Indian Ocean. (Image courtesy of NASA/JPL-Caltech.)*

Figure 96–2 *Uncle Milton Moon in My Room.*

Figure 96–3   *Functioning as an astronomical night light, Moon in My Room is also an accurate representation of the lunar surface.*

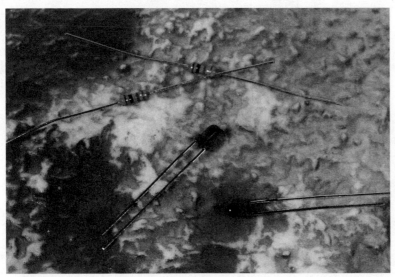

Figure 96–4   *Plot the landing sites for the Apollo mission series. Use red LEDs, along with some current limiting resistors, for marking each Apollo lunar base point. You can learn more about the Apollo missions at the NASA Historical Archives for the Apollo missions (www.hq.nasa.gov/ office/pao/ History/apollo.html).*

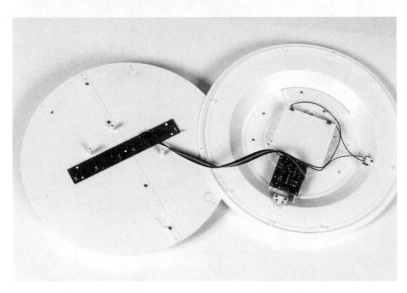

Figure 96–5   *Inside Moon in My Room is ample space for holding your Apollo LEDs.*

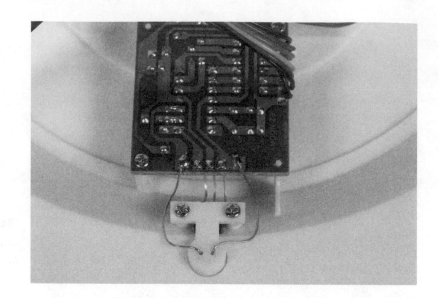

Figure 96–6 *You can draw your power from the main circuit board.*

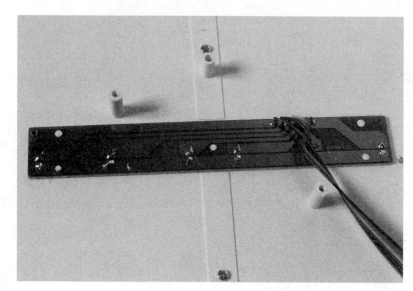

Figure 96–7 *You could try to piggyback on the lunar "phase" LEDs already built-in to Moon in My Room.*

Project 97. How to Show the Phases of the Moon

What You Need:

- Uncle Milton Moon in My Room

Resources:

- Uncle Milton Toys—www.unclemilton.com

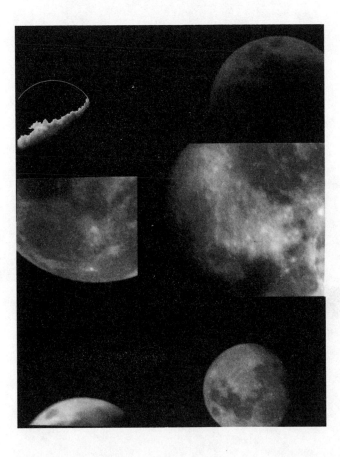

Figure 97–1 *Make your own moon shots. It's easy. Just refer to some of the projects in Chapter 7 for inspiration and guidance.*

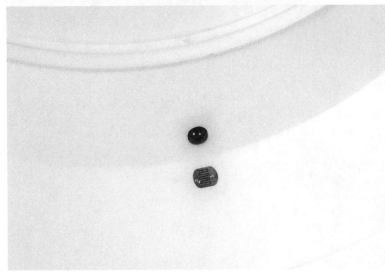

Figure 97–2 *An IR receiver for enabling the handheld controller and a photoresistor for making your Moon in My Room "light up" at night, just like the real thing.*

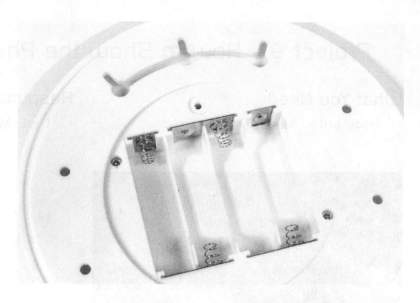

Figure 97–3   *Four batteries and a special mount for hanging your moon on your wall.*

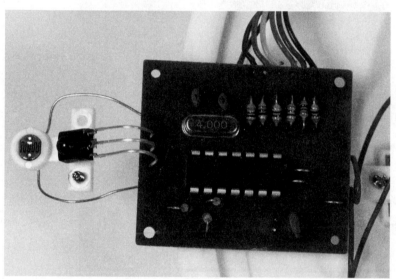

Figure 97–4   *An internal circuit board supports the IR receiver and photoresistor.*

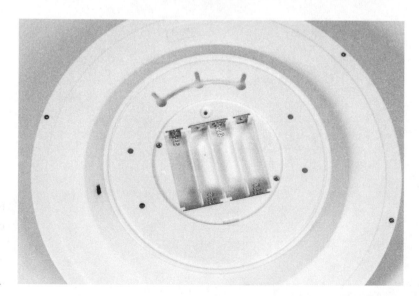

Figure 97–5   *A lot of screws must be removed for exposing this moon's phases.*

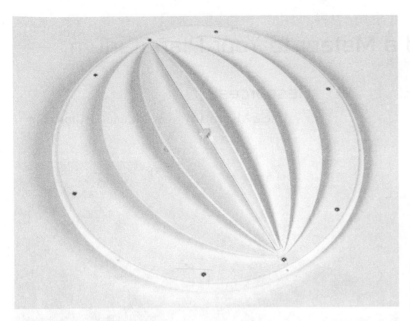

Figure 97–6 *A series of innovative illuminated elliptical wedges give this moon its phases.*

Project 98. Build a Home Planetarium

What You Need:

- Flashlight
- Foil
- Pin

Use the Uncle Milton Star Theater 2 as your reference for building your own home planetarium. Carefully, lay a sheet of foil over the Star Theater 2 projector. Now, prick a small hole in the foil where ever a star appears on the projector. Remove the foil, darken the room's lights, and shine a flashlight through the back of the foil. Light shining through the foil's holes will give a rough approximation of the night sky.

Figure 98–1 *The Uncle Milton Star Theater 2 is a great start for building your own home planetarium.*

Project 99. Add a Meteor to Your Planetarium

What You Need:

- Uncle Milton Star Theater 2

Resources:

- Uncle Milton Toys—www.unclemilton.com

Figure 99–1 *Star Theater 2 includes a special projection wand that can display actual photographs (the two discs located at the bottom of the figure) of meteors inside your planetarium.*

Figure 99–2 *You can remove the Star Theater 2 meteor maker for projecting actual meteor photographs inside your own home planetarium.*

Project 100. Take Your Planetarium Outdoors for a Real Star Show

What You Need:

- Green laser pointer
- Uncle Milton Star Theater 2

Resources:

- Uncle Milton Toys—www.unclemilton.com

Figure 100–1 *Take your Star Theater 2 outside to help you identify the constellations. The glow from previously charged (illuminated) phosphorescent stars can help you visualize the sky above you.*

Chapter Eleven

OK, Go Make Your Own Heavenly Bodies

The report on Unidentified Aerial Phenomena (UAP) by the British Ministry of Defence (sic) [Ministry of Defence Unidentified Aerial Phenomena (UAP) in the UK Air Defence Region (December 2000; Scientific & Technical Memorandum No. 55/2/00)] couldn't have said it more succinctly:

> "9. Based on all the available evidence remaining in the Department (reported over the last 30 years), the information studied, either separately or corporately contained in UAP reports, leads to the conclusion that it does not have any significant Defence Intelligence value. However, the Study has uncovered a number of technological issues that may be of potential defence interest.

> "10. Causes of UAP Reports. In the absence of any evidence to the contrary, the key UAP report findings are:

> [Excerpt edited.]

> "Further:

> "No evidence exists to associate the phenomena with any particular nation.

> "No evidence exists to suggest that the phenomena seen are hostile or under any type of control, other than that of natural physical forces.

> "Evidence suggests that meteors and their well-known effects and, possibly some other less-known effects, are responsible for some UAP."

Well, that says it; there's no such thing as UAP or Unidentified Flying Objects (UFOs) piloted by aliens flying around Earth. Or, are there? With over 700+ sightings reported to the Ministry of Defence during 1978 alone, there are more than a few folks seeing meteors, aren't there?

What is needed is good solid evidence. It's up to evil geniuses like you to find that evidence. So let's get to work.

Resources:

- Ministry of Defence Unidentified Aerial Phenomena (UAP) in the UK Air Defence Region—www.mod.uk/DefenceInternet/ FreedomOfInformation/PublicationScheme/ SearchPublicationScheme/Unidentified AerialPhenomenauapInTheUkAirDefence Region.htm

- *SCHWA® World Operations Manual*, by The SCHWA Corporation (Chronicle Books, 1997)

- *The U.F.O. Hunter's Handbook: A Field Guide to the Paranormal,* by Caroline Tiger (Book Soup Publishing, Inc., 2001)

- Unidentified Flying Objects and Air Force Project Blue Book—www.af.mil/factsheets/ factsheet.asp?id=188

- Unidentified Flying Objects: Project BLUE BOOK—www.archives.gov/foia/ufos.html

What You Need:

- Some paper plates
- Film camera (stay away from digital cameras; the "man" will call you a fake)
- Monofilament line
- Bits of metal foil
- Poor light

Figure 101–1 *Gone are the days of having to rely on nylon thread for holding your hoax spacecraft in position while you snapped an award-winning UFO photo. Start with a believable foreground image.*

Figure 101–2 *Even though you will be using Adobe Photoshop for mastering your hoax, you still need to build a model for your extraterrestrial sighting.*

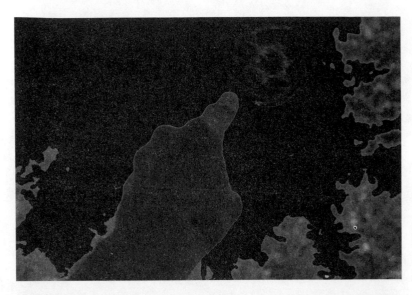

Figure 101–3   *By combing Figures 101-1 and 101-2, you can get this UFO encounter. The solarized film helped to make this fake night image look believable.*

Figure 101–4   *Try different angles for your foreground image.*

Figure 101–5   *By building a real fake spacecraft model, you can easily photograph different views for subsequent forgery.*

Figure 101–6 *Sometimes UFOs happen during the daylight. Merging Figures 101-4 and 101-5 together could help scramble USAF fighters for intercepting this bogus target.*

Figure 101–7 *It takes a lot of practice to sell a good con.*

Figure 101–8 *Subtle photographs of your fake spacecraft model can make the final photograph believable.*

Figure 101–9   *Contact the local newspapers. I think we've spotted a real UFO this time.*

Appendix A

Resources

All of the Web addresses listed in the text have been collected into this one convenient location:

- Academy—www.academyhobby.com
- Aiptek—www.aiptek.com
- AIRNow—www.airnow.gov/index.cfm?action=airnow.national
- American Astronomical Society—www.aas.org/
- Apogee Rockets—www.apogeerockets.com
- The Astronomical Journal—www.journals.uchicago.edu/AJ/
- Astronomy & Astrophysics—www.aanda.org/
- Astronomy Technologies, Inc.—www.astronomytechnologies.com
- BOINC: Compute for Science—boinc.berkeley.edu
- Camera hacking forum featuring John Maushammer—www.camerahacking.com
- Canon U.S.A.—www.usa.canon.com
- Celestia—www.shatters.net/celestia
- C. Crane Company, Inc.—www.ccrane.com
- Earth Science Information Centers—geography.usgs.gov/esic/esic_index.html
- An Empty Spacesuit Becomes an Orbital Experiment (NASA)—www.nasa.gov/mission_pages/station/expeditions/expedition12/26jan_suitsat.html
- Estes Rockets—www.estesrockets.com
- European Southern Observatory—www.eso.org/

- Google Earth—earth.google.com
- Great Red Spot Astronomy Products—www.greatredspot.com
- Harvard SETI—seti.harvard.edu/seti/
- How to Use a Compass—www.learn-orienteering.org
- The Hubble Space Telescope Data Archive—www.adass.org/adass/proceedings/adass94/bornek.html
- ICARUS, International Journal of Solar System Studies—icarus.cornell.edu/
- Imaginova—www.imaginova.com
- International Orienteering Federation—www.orienteering.org
- Jetex Organization—jetex.org
- JoinMax Digital Tech. Ltd.—www.robotplayer.com
- JoinMax RobotEXP—www.robotexp.com/
- Konus—www.konus.com
- LEGO Group—www.lego.com
- LiveScience.com—www.livescience.com
- Astronaut Bio: Shannon W. Lucid (188 days)—www.jsc.nasa.gov/Bios/htmlbios/lucid.html
- MacDEM—www.treeswallow.com/macdem
- Magellan Navigation, Inc.—www.magellangps.com
- MapMart—www.mapmart.com
- Maps a la carte, Inc.—www.topozone.com
- Maptech—mapserver.maptech.com

- Mars Exploration Rover Mission—marsrovers.nasa.gov/home/

- Meade Instruments Corporation—www.meade.com

- MICRODEM—www.usna.edu/Users/oceano/pguth/website/microdem.htm

- Microsoft TerraServer Imagery—terraserver-usa.com

- MINDSTORMS—www.mindstorms.com

- Ministry of Defence Unidentified Aerial Phenomena (UAP) in the UK Air Defence Region—www.mod.uk/DefenceInternet/FreedomOfInformation/PublicationScheme/SearchPublicationScheme/UnidentifiedAerialPhenomenauapInTheUkAirDefenceRegion.htm

- Multimission Archive at Space Telescope (MAST) HST—archive.stsci.edu/hst/

- National Aeronautics and Space Administration—www.nasa.gov

- NASA Astrobiology—astrobiology.arc.nasa.gov

- NASA Astronaut Breaks U.S. Space Endurance Record (375 cumulative days in space) —www.nasa.gov/home/hqnews/2003/dec/HQ_03400_Foale_record.html

- NASA Historical archives for the Apollo missions—www.hq.nasa.gov/office/pao/History/apollo.html

- NASA Radio JOVE Project—radiojove.gsfc.nasa.gov

- NASA Robotics: The Robotics Alliance Project—robotics.nasa.gov

- NASA Star Count, Student Observation Network—www.nasa.gov/audience/foreducators/starcount/home/index.html

- National Association of Rocketry—www.nar.org

- National Atlas of the United States, March 5, 2003—nationalatlas.gov

- National Institute of Standards and Technology—tf.nist.gov/general/publications.htm

- National Space Science Data Center—nssdc.gsfc.nasa.gov/

- Nikon, Inc.—www.nikonusa.com

- Night Owl Optics—www.nightowloptics.com

- Polar Heart Rate Monitors—www.polarusa.com

- Biographies of USSR/CIS Cosmonauts Polyakov, Valeri (437 days from 1994–1995 aboard Soyuz TM-18)—www.spacefacts.de/bios/cosmonauts/english/polyakov_valeri.htm

- Quest Aerospace—www.questaerospace.com

- Rapier Pardubice—www.rapier.cz/index.htm

- Reiter Breaks European Space Endurance Record (209 days in orbit)—www.esa.int/esaCP/SEME93JZBQE_index_0.html

- *SCHWA World Operations Manual*, by The SCHWA Corporation (Chronicle Books, 1997)

- Search for Extraterrestrial Intelligence: The Planetary Society—www.planetary.org/explore/topics/seti/

- SETI@home—setiathome.berkeley.edu

- SETI Institute—www.seti.org/site/pp.asp?c=ktJ2J9MMIsE&b=178025

- SETI League—www.setileague.org/

- Silva USA—www.silvausa.com

- Smithsonian Astrophysical Observatory Star Catalog—heasarc.gsfc.nasa.gov/W3Browse/star-catalog/sao.html

- Space.com—www.space.com

- The Space Place—spaceplace.jpl.nasa.gov/en/kids/

- Squadron Shop—www.squadron.com

- Starry Night Store—www.starrynight.com

- Stellarium—www.stellarium.org

- SuitSat-1 RS0RS—www.amsat.org/amsat-new/ articles/BauerSuitsat/index.php

- Suunto Compasses—www.suunto.com

- Tele Vue Optics, Inc.—www.televue.com

- Uncle Milton Toys—www.unclemilton.com

- *The U.F.O. Hunter's Handbook: A Field Guide to the Paranormal,* by Caroline Tiger (Book Soup Publishing, Inc., 2001)

- Unidentified Flying Objects and Air Force Project Blue Book—www.af.mil/

- factsheets/factsheet.asp?id=188

- Unidentified Flying Objects: Project BLUE BOOK—www.archives.gov/foia/ufos.html

- U.S. Geological Survey—www.usgs.gov

- USGS Digital Raster Graphics—topomaps.usgs.gov/drg/

- Vivitar—www.vivitar.com

- William Optics—www.williamoptics.com

- Zhumell, Inc.—www.zhumell.com

Appendix B

A Celestial Almanac

This is a listing of significant 2007 astronomical events:

January 2007	Quadrantids Meteor Shower
1/3/2007	Full Moon
1/25/2007	First Quarter Moon
2/2/2007	Full Moon
2/10/2007	Saturn Opposition
2/17/2007	New Moon
2/23/2007	Mercury Inferior Conjunction
2/24/2007	First Quarter Moon
3/3/2007	Total Lunar Eclipse
3/3/2007	Full Moon
3/12/2007	Last Quarter Moon
3/19/2007	Partial Solar Eclipse
3/21/2007	Vernal Equinox
3/25/2007	First Quarter Moon
April 2007	Lyrids Meteor Shower
4/2/2007	Full Moon
4/17/2007	New Moon
4/24/2007	First Quarter Moon
May 2007	Eta Aquarids Meteor Shower
5/2/2007	Full Moon
5/10/2007	Last Quarter Moon
5/23/2007	First Quarter Moon
6/1/2007	Full Moon
6/5/2007	Jupiter Opposition
6/15/2007	New Moon
6/19/2007	Pluto Opposition
6/21/2007	Summer Solstice
6/22/2007	First Quarter Moon
6/30/2007	Full Moon
July 2007	Southern Delta Aquarids Meteor Shower
7/7/2007	Last Quarter Moon
7/14/2007	New Moon
7/30/2007	Full Moon
July/August 2007	Alpha Capricornids Meteor Shower
August 2007	Perseids Meteor Shower
8/5/2007	Last Quarter Moon
8/12/2007	New Moon
8/13/2007	Neptune Opposition
8/15/2007	Mercury Superior Conjunction
8/20/2007	First Quarter Moon
8/21/2007	Saturn Conjunction
8/28/2007	Total Lunar Eclipse
8/28/2007	Full Moon
9/4/2007	Last Quarter Moon
9/11/2007	Partial Solar Eclipse
9/11/2007	New Moon
9/19/2007	First Quarter Moon
9/23/2007	Autumnal Equinox
9/26/2007	Full Moon

October 2007	Orionids Meteor Shower	December 2007	Geminids Meteor Shower
10/3/2007	Last Quarter Moon	December 2007	Ursids Meteor Shower
10/19/2007	First Quarter Moon	12/17/2007	First Quarter Moon
10/26/2007	Full Moon	12/21/2007	Pluto Conjunction
November 2007	Taurids Meteor Shower	12/22/2007	Winter Solstice
November 2007	Leonids Meteor Shower	12/23/2007	Jupiter Conjunction
11/1/2007	Last Quarter Moon	12/24/2007	Full Moon
11/17/2007	First Quarter Moon	12/24/2007	Mars Opposition
11/24/2007	Full Moon		

Index

Index